Nematodes in Soil Ecosystems

Nematodes in Soil Ecosystems

Edited by Diana W. Freckman

Foreword by J. A. Wallwork

 University of Texas Press, Austin

Contents

LIST OF CONTRIBUTORS

Richard V. Anderson, Department of Biological Sciences,
Western Illinois University, Macomb, Illinois 61455

D. C. Coleman , Department of Zoology/Entomology and Natural
Resource Ecology Laboratory, Colorado State University,
Ft. Collins, Colorado 80523

D. A. Crossley, Jr., Institute of Ecology, University of
Georgia, Athens, Georgia 30602

Larry Duncan, Department of Nematology, University of
California, Riverside, California 92521

N. Z. Elkins, Department of Biology, New Mexico State
University, Las Cruces, New Mexico 88003

Howard Ferris, Department of Nematology, University of
California, Riverside, California 92521

J. M. Ferris, Department of Entomology, Purdue University,
West Lafayette, Indiana 47907

V. R. Ferris, Department of Entomology, Purdue University,
West Lafayette, Indiana 47907

Diana W. Freckman, Department of Nematology, University of
California, Riverside, California 92521

P. B. Goodell, University of California Cooperative Extension,
P.O. Box 2509, Bakersfield, California 93308

Thomas B. Kirchner, Natural Resource Ecology Laboratory,
Colorado State University, Ft. Collins, Colorado 80523

R. McSorley, University of Florida, Agricultural Research
and Education Center, Homestead, Florida 33031

L. W. Parker, Department of Biology, New Mexico State
University, Las Cruces, New Mexico 88003

P. F. Santos, Departmento de Ecologia, Instituto de Biocencias,
UNESP Campus de Rio Claro, Caixa postal 178, 13.500 Rio
Claro Sao Paulo, Brazil

B. R. Stinner, Institute of Ecology, University of Georgia
Athens, Georgia 30602

J. A. Trofymow, Natural Resource Ecology Laboratory, Colorado
State University, Ft. Collins, Colorado 80523

W. G. Whitford, Department of Biology, New Mexico State
University, Las Cruces, New Mexico 88003

G. W. Yeates, Soil Bureau, DSIR, Lower Hutt, New Zealand

FOREWORD

In view of their abundance and wide distribution, it is surprising that so little is known about the biology and, more
particularly, the ecology of free-living soil nematodes.
There are, of course, reasons for this. As members of the
soil-water fauna they can only be recovered satisfactorily
from soil samples by "wet" extraction methods. Soil zool-
ogists, in the main, tend to favor "dry" extractors since the
rewards are great in terms of the fauna diversity which can
be recovered by these methods. Consequently, the nematode
component of this diversity remains very largely overlooked.
Again, soil nematodes are minute in size and species' identification requires the assistance of the specialist taxonomist.
The small body size of soil nematodes is more than compensated
for by their high population densities, and this fact should
act as an incentive rather than a deterrent to their study.
However, taxonomic obstacles remain to be overcome.

This last problem, the taxonomic one, has been solved or
circumvented in part by the recognition of different trophic
groups, or "feeding guilds," among soil nematodes which can
form the basis for an ecological classification. Such a
scheme distinguishes between phytophages, predators, fungivores, microbial-feeders and omnivores, although the distinc-
tion between these groupings is not always clearly drawn.
Nevertheless, this approach emphasizes the basic functional
roles of nematode species in the soil community, and provides
a starting point for the formulation of ecological questions.
For example, the relative proportions of different guilds
within a local nematode fauna may reflect the availability of

a particular food resource and thereby indicate the precise
pathway along which energy and materials are flowing. Productivity studies become more meaningful when they are directed at the guild, rather than at individual species or
entire nematode faunas. Ultimately these considerations can
lead to an understanding of the nature of the decomposition
process occurring in soils, and the extent to which nematodes
are involved in this process.

    Given the impetus of this approach, ecological studies on
soil nematodes are now proceeding along a number of broad
fronts. The association of nematodes with such bacteria-rich
microhabitats as the rhizosphere focuses attention on animal/
microbial interactions at a first-order level. The possibility exists that nematodes may regulate rates of organic decomposition indirectly through their depredations on the
bacterial decomposers. Such activities will bring them into
direct competition with other members of the microbial feeding fuild: protozoans, certain prostigmatid and cryptostigma-
tid mites and Collembola. The nature and intensity of these
competitive interactions are largely unknown. Another dimension is added to this picture by the fact that some prostigmatid and mesostigmatid mites are known to prey directly on
soil nematodes. Perturbation experiments carried out in the
field in which these predators have been selectively elimi-
nated have revealed a slowing down of the rate of organic
decomposition. The inference to be drawn from this is that
a relaxation of predator pressure has allowed an increase in
nematode densities and an intensification of their bacteriophagous activities. Therefore, it seems possible that under
natural conditions predatory mites may regulate nematode
densities and positively promote the decomposition of organic
material. This interpretation may be too simplistic, partic-

ularly since one can visualize similar interactions occurring between nematodes, fungi and actinomycetes. The possibility of competition between bacteria, fungi and actinomycetes for organic substrates also needs consideration. Each new parameter brings with it an added degree of complexity; a new set of cross-links is introduced into the food web system.

How, then, can we unravel this complexity? A good starting point, surely, is to select a relatively simple soil system in which microfloral/nematode/predator interactions can be identified as important determinants of the type and rate of organic decomposition. Recent work carried out in the southwestern United States strongly suggests that desert soils provide such a system. Here, nematode densities and diversity are comparable with those of cool, moist temperate forest soils, and while interactions between the soil microflora and fauna are not quite as simple as perhaps was originally supposed, these interactions can be identified and analyzed by stepwise regression techniques, as we will see later in this book. In these desert systems, perturbation experiments such as those mentioned earlier can be very useful for the identification of interactions, and this will also become apparent later.

Apart from their relationships with other biota, soil nematodes are intrinsically interesting, and those of arid soils particularly so. The successful colonization of these soils by animals which are essentially aquatic in habit may be attributable, at least in part, to their ability to assume an anhydrobiotic state during periods of drought. This exciting area of study is gaining more and more attention, particularly since it raises the possibility that other members of the invertebrate fauna of arid soils may respond in a similar manner to moisture stress. Here is a good example of the way in which studies on soil nematodes are leading the way into new fields of physiological ecology which may have far-reaching applications.

The contributors to this volume are all actively engaged in this vigorous quest for knowledge concerning the role of nematodes in soil ecosystems. What follows is eloquent testimony to their efforts and an accurate reflection of the progress being made in this field.

J. A. Wallwork
Westfield College
University of London

PREFACE

Although nematodes are as numerous as arthropods, the contribution of the nematode community to the soil ecosystem is just
beginning to be assessed. The paucity of available information on the role of the total nematode community in the soil
may be because historically it has been studied as several independent areas of research. In past centuries, the study of
soil nematodes has been delegated to the agricultural nematologists due to the direct relationship between losses in
crop yield and the abundance of plant parasitic nematodes. To
these nematologists, we owe our knowledge of methodology, tax-
onomy, biology, and interactions of mainly phytoparasitic
nematodes. More recently, ecologists have been interested in
the structure and function of the total soil nematode community, which includes not only the primary consumers of roots,
but bacterivorous, fungivorous, predaceous, and omnivorous
nematodes and their relation to other soil microflora and
fauna. Because of the ecologist's interest in the flow of
energy and cycling of nutrients, they have contributed areas
of knowledge previously unexamined by agricultural nematologists. Investigators from the ecological and agricultural
areas seem unaware of each other's work. To stimulate communication between the soil ecologists and agricultural
nematologists, a symposium was held in the summer of 1980 at
the annual meeting of the American Institute of Biological
Sciences in Tucson, Arizona. This book contains papers presented at that symposium, some of which are meant to encourage
thought and others to offer data and information to aid us in
our understanding and future assessment of the nematode's
role in various soil ecosystems.

The presentations at the symposium were divided into distinct areas, primary consumption, decomposition, and synthesis and validation, as are the papers in this book. It is hoped that in reading these papers, one can appreciate the necessity for good sampling techniques (Goodell) which will influence quantitative estimates of soil nematodes (Freckman) in the ecosystem. Primary production as affected by the different life habits and energy requirements of phytoparasites is examined by Ferris, and the effect of interspecific nematode interactions on primary production is discussed by Duncan and Ferris. Phytoparasites and other members of the nematode community are discussed in a no-till agroecosystem (Stinner and Crossley). The present state of knowledge of nematodes and decomposition is summarized by Yeates and Coleman and in desert soils by Whitford et al. A more specific study of nematodes and their role in cellulose and chitin decomposition is discussed by Trofymow and Coleman. Information derived from various studies has been used to synthesize and validate two different models; the prediction of population fluctuations of a primary consumer on corn roots (McSorley et al.), and the prediction of population dynamics and carbon flow in bacteriophagic nematodes (Anderson and Kirchner).

The goal of this volume then is to further communication between soil ecologists and nematologists and to stimulate more studies of this large and diverse group of soil fauna.

Diana W. Freckman
Riverside, CA

Part I
Primary Consumption

THE ROLE OF NEMATODES AS PRIMARY CONSUMERS

Howard Ferris

Plant-parasitic nematodes are generally studied as causal agents of plant damage. Analysis of the role of nematodes as primary consumers forces consideration of the routes through which nematodes supply energy into the soil ecosystem. Some assumptions and generalizations were necessary. According to Sohlenius (1980), nematode densities range from 3.5 million per m^2 in tundra to 9 million per m^2 in a temperate grassland. Fewer nematodes were found in a desert ecosystem and, somewhat surprisingly, in a tropical forest ecosystem. However, these studies were performed by different researchers, and techniques resulting in different extraction efficiencies were employed. Plant-parasitic nematodes averaged about 21% of the nematodes in undisturbed ecosystems and up to 35% in disturbed eco- systems. Nematode communities show greater seasonal variation relative to cropping and tillage practices in disturbed eco- systems.

CHARACTERISTICS OF PLANT-PARASITIC NEMATODES
Plant-parasitic nematodes have the general characteristics of being equipped with a protrusible stylet and associated gland- ular structures involved in the predigestion of cell contents, and a muscular pharyngeal pump to facilitate ingestion. They vary in the location and sophistication of their association with the root. The nature of the association is a determinant in the rate of energy flow through the nematode system. Nema- todes can be characterized according to their feeding habits. Sedentary endoparasites, such as the root-knot nematode, Meloidogyne spp., feed entirely within the root system. They

may promote the development of specialized feeding sites such
as syncytia or otherwise enlarged cells. These metabolically
active cells act as a sink for photosynthates and channel them
into the nematode system. Many endoparasitic nematodes become
immotile during their life cycle and may become enlarged, functioning essentially as egg production machines. Another group
of endoparasites remains migratory in the root tissues and does
not stimulate the production of specialized feeding sites.
Ectoparasitic nematodes generally remain motile and migratory
on the outside of the root. They may feed from superficial
root cells or, if equipped with longer stylets, from deeper
root tissues. Somewhat intermediate are sedentary semi-endoparasitic nematodes, such as cyst nematodes which feed on
specialized cells in the vascular tissue, with most of the body
outside the root. Migratory endoparasites feed within the
root system but remain motile throughout their life cycle so
that, as conditions deteriorate in one part of the root, they
are able to migrate to new feeding areas. Generally they feed
from cells in the cortical region of the root. Sedentary nematodes usually deposit their eggs in masses, often embedded in
a gelatinous matrix, while migratory nematodes usually deposit
their eggs individually. The life style and the reproductive
potential of the nematodes indicates the amount of energy which
will be funneled into the soil ecosystem.

The anatomical and physiological changes caused by the root-
knot nematode may result in considerable damage to the root.
Consequently, plant vigor is reduced, less food is available
to the nematode, and there is a logistic-type depression of
the rate of nematode population development. The root-knot
nematode enters the root as a second-stage infective larva and
becomes oriented within the vascular system, inciting the
development of feeding syncytia, and maturing through a series

of larval stages to a swollen egg-laying female. The nematode
then functions as a machine pumping out eggs, merely moving
its head between syncytia, resulting in a low energy cost
associated with feeding. A sedentary semi-endoparasitic nematode such as Rotylenchulus reniformis feeds with its head deep
within the root tissues, in the region of the pericycle. As
with the root-knot nematode most of the food intake of the
adult nematode will be channeled into the production of eggs,
rather than to energy used for movement and food seeking.
Because there is less root disruption than with the root-knot
nematode, there is less damage per individual to the root
system and therefore less of a reciprocal suppressant effect
on the nematode population development. A migratory endoparasite within the cortical tissues deposits eggs individually
as it moves among feeding sites. Presumably more energy is
expended in this feeding processs than with sedentary endoparasites. However, the amount of damage per nematode and its
influences on the plant is probably less, so that the rate of
population development is less affected by plant damage.
Migratory ectoparasites remain on the outside of the root,
although they may be feeding deep within the root tissues, and
cause considerable disruption of the root system in some cases.
They remain active and may follow the growing root tip.
Obviously, the energy expended in feeding and movement is
considerably greater than for sedentary nematodes.

ENERGY FLOW THROUGH THE SYSTEM
It is necessary to consider the life style and feeding habit
of nematodes to discuss the partitioning of energy intake and
the efficiency of utilization. These factors dictate the
amount of energy passing to the next trophic level of the soil
ecosystem rather than being dissipated as heat. If we consider

the nematode system as an energy tank (Fig. 1) with inlets and
outlets controlled by valves which regulate energy flow, it
is possible to subjectively partition the relative amount of
energy flowing through each valve according to life habit of
the nematode.

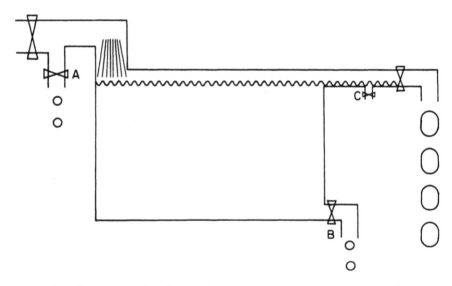

Fig. 1.   Energy tank of a sedentary endoparasitic nematode
system (e.g. Meloidogyne sp.). A) Metabolic cost of food
intake; B) Metabolic cost of maintenance and respiration;
C) Metabolic cost of reproduction.

Valve A governs the amount of energy being used in the uptake
of food, probably less for a sedentary endoparasite with
specialized feeding sites than for a migratory ectoparasite
or migratory endoparasite. The actual amount of food flowing
into the system is a function of the relationship which the
nematode establishes with its host. The volume of food ingested is much greater for nematodes supplied by specialized

feeding sites than for those feeding from unaltered host
cells.

Valve B represents respiration and maintenance, that is,
the metabolic cost of keeping the existing biomass alive and
functional. The amount flowing through valve B is a function
of the environmental conditions, body size and activity of the
nematode at that time. Once the body tank has filled and the
nematode has reached its adult size, priorities of the nematode
change with maturity, growth ceases, and egg production commences. There is a continuing respiratory cost for the established adult tissues, and also a food intake cost. The balance
of available energy can be channeled into egg production.
There may also be a reproductive cost (Valve C) if mating behavior is involved, although many plant-parasitic nematodes
have resorted to a parthenogenetic strategy. Under conditions
of stress, some parthenogenetic nematodes produce non-essential
males rather than females, increasing the flow through their
reproductive cost valve (C) as energy flows into male production rather than eggs. In sexually reproducing nematodes,
there is a normal reproductive cost of energy channeled into
male production as males are required for the system. Since
males do not produce eggs, their production lowers the rate
of population increase.

The root-knot nematode (Fig. 1) has a large food intake at
relatively low feeding cost, fairly high metabolic cost associated with the body volume, but little energy cost due to
movement, and no reproductive cost of male production. The
balance of the energy input into the system is channeled into
the production of eggs. These nematodes characteristically
have a high egg output. As an extreme, consider a migratory
ectoparasite such as Paratylenchus (Fig. 2). This nematode
feeds in the vicinity of the root tip, bracing against soil

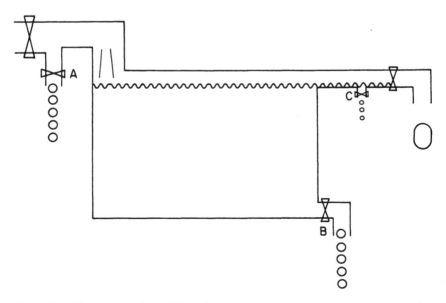

Fig. 2.   Energy tank model of migratory ectoparasitic nematode
system (e.g. <u>Paratylenchus</u> sp.).   A) Metabolic cost of food
intake; B) Metabolic cost of maintenance and respiration; C)
Metabolic cost of reproduction.

particles to insert its stylet into the root, resulting .in a
rather high energy cost associated with feeding.   Further,
because the nematode is feeding on unaltered cells, which do
not supply food at the same rate as adapted syncytia, the
volume flowing into the system is lower.   Respiratory costs
are probably higher than for a sedentary nematode because of
the activity level.   Also, these nematodes may be sexually
reproducing, resulting in partitioning of the energy into the
production of males.   The net result is lower egg production.

Applying caloric values to these perceptions of the life
styles and energy dynamics of nematodes as primary consumers
provides some indication of the flow of energy they contribute
to the soil ecosystem. Multiplying the body size of various
nematode types by published values for caloric equivalents
(de Soyza 1973) indicates the energy cost of the biomass. The
number of life cycles they pass through in a year will dictate
their productivity. The period of productivity occurs during
the time of the year that the nematodes are metabolically
active or at least reproductively active, probably about six
months in temperate regions. From published data on life
cycle lengths (Norton 1978), the number of generations in six
months for nematodes of different life habit ranges from 3 to
8 (Table 1). The number of eggs produced per female per
generation (Norton 1978) allows calculation of total annual
egg productivity for each life style of nematode. In these
calculations the assumption is made that egg mortality may be
as high as 90% at each generation, and that nematodes are increasing without food or space limitations (Table 1). This
high mortality, indicated by population data from sampling
studies, may occur during the egg stage from harsh conditions,
parasitism or predation, or may reflect failure of larval
nematodes to locate a host root and establish a feeding site.
The result is an energy supply from the previous generation
liberated into the remainder of the soil ecosystem. Note the
estimate of 2 million root-knot eggs per year, even though
there are only 3 generations, because of the high egg production
rate. There are fewer eggs for other nematodes that have more
generations because of the lower egg production. In general
the highest production per year occurs with sedentary nematodes
and decreases in migratory forms because of the lower food intake and high energy cost associated with the life habit.

TABLE 1. Estimates of annual caloric production of plant-parasitic nematodes with different feeding habits. Assumptions include six months of metabolic activity and 90% egg mortality per generation. NOTE--these estimates are made for the purpose of discussion and have not been experimentally verified.

| | PRODUCTION | | | | SUPPLY TO SYSTEM (MORTALITY) | | |
	cal/ individual (x1000)	generations	eggs/ female	nematodes/ year (x.001)	cal/year nematodes	cal/year egg mortality	total supply to system
Meloidogyne (sedentary endoparasite)	21.45	3	1344	2427.7	52074	903	52977
Rotylenchulus (sedentary ectoparasite)	4.05	5	120	248.8	1008	93	1101
Aphelenchoides (migratory endoparasite)	0.49	8	48	281.8	137	105	242
Radopholus (migratory endoparasite)	0.47	5	72	19.3	9	7	16
Paratylenchus (migratory ectoparasite)	0.08	4	138	36.3	3	14	17

The energy supply to the soil ecosystem from juvenile and
adult nematodes is determined by multiplying the biomass by
their caloric values. About 52,000 calories are produced from
the large bodied root-knot nematode assuming that the adults
die at the end of each generation (Table 1). Because of the
large biomass of adult females, the contribution of the egg
mortality is relatively low, even though many eggs are involved. In nematodes with smaller female size, the contribution of egg mortality is relatively greater. In migratory
ectoparasites, which have a high energy life style, the energy
contribution of egg mortality to subsequent trophic groups of
the soil ecosystem may be 3 or 4 times greater than the contribution of the adult mortality (Table 1). A probable bias
in this rationale is that migratory or sedentary endoparasities,
once they have established an infection site or are within the
root system, may be isolated from predators and parasites.
However, ectoparasites remain in the soil environment and are
more likely to be exposed to parasites and predation. Con-
ceivably, the mortality of juvenile forms of ectoparasites is
somewhat greater, resulting in a larger relative contribution
to the energy flow through the soil ecosystem than from adults.

The metabolic costs of production includes a food intake
efficiency cost (Fig. 1, valve A) relative to the amount of
energy expended in ingestion. Cost in terms of growth and
respiration is a function of body size and life habit, as
well as environmental conditions. Metabolic costs of egg
production are related to the number of eggs produced. There
are respiration costs after the nematode achieves adulthood
and is producing eggs. The cost for sedentary endoparasites
is higher than for ectoparasites due to their more efficient
channeling of energy. The approximated data allow estimation
of the production efficiency of root-knot nematodes at about

25 to 30% and of migratory ectoparasites such as <u>Paratylenchus</u> about 18%. Note that all data generated in this presentation are hypothetical estimations based on perceptions of the life cycle and life habits of plant-parasitic nematodes. They have not been generated by experimentation. In generating the data, exponential growth of nematode populations was assumed. In fact, population studies have shown that the magnitude of the initial population will influence the growth of the plant and, reciprocally, the amount of food available to the nematode. The damage to the plant per nematode influences the production of the nematode system, therefore, a necessary sophistication of the above rationale would be to consider the carrying capacity of the plant as a logistic feedback in hypothesizing energy flow through the primary consumers.

In summary, the contribution of plant-parasitic nematodes to the energy supply of a soil ecosystem involves not only the numbers of nematodes and their weights and sizes, but also their life styles, feeding habits, multiplication rates, and reproductive strategies. Other determinant factors include their damaging effects on the host, the population age structure, and high age-specific mortality such as high rates of egg and preparasitic larval mortality. Much of the biology for calculating energy flow in plant-parasitic nematodes is available. It merely remains to ask the appropriate questions to extract the information.

LITERATURE CITED

Sohlenius, B. 1980. Abundance, biomass and contribution to energy flow by soil nematodes in terrestrial ecosystems. Oikos 34:186-194.

Norton, D. C. 1978. Ecology of plant-parasitic nematodes. John Wiley, Inc., New York, New York, U.S.A.

de Soyza, K. 1973. Energetics of _Aphelenchus_ _avenae_ in
    monoxenic culture. Proceedings of the Helminthological
    Society of Washington 40:1-10.

NEMATODES IN NO-TILLAGE AGROECOSYSTEMS

B. R. Stinner and D. A. Crossley, Jr.

INTRODUCTION

Nematodes are abundant members of the soil fauna in a variety
of ecosystem types. Information on the effects of nematodes
on biological processes in the soil has been slow to develop
(Sohlenius 1980), possibly due to discrepancies in sampling
techniques and the large variation inherent in nematode population sizes. Recently, innovative research with manipula-
tory systems such as microcosms (Anderson et al. 1978,
Coleman et al. 1977) has provided information on the significance of nematodes in the decomposition process. Still, the
bulk of research reported on nematodes in soil concerns agricultural systems, where plant parasitic nematodes can be
responsible for crop losses. Nematodes certainly have other
effects in agronomic systems, such as influencing mineralization rates (Wasilewska et al. 1981), but these effects are
largely unknown (Yeates 1979). Effective management of
economically important nematodes may depend ultimately upon
knowledge of their relation to soil processes. For example,
control of plant parasitic nematodes by additions of organic
amendments is well documented but the mechanism of control is
not yet understood (Mashkoor et al. 1977).

Studies of nematodes in lightly managed systems (Yeates
1979) should contribute information on the impact of nematode
activities on soil biological processes. Such systems are
intermediate between the severely modified cultivated agronomic fields and the complex natural, undisturbed soil systems.
No-tillage and minimum tillage systems may be regarded as
lightly managed from the viewpoint of minimum soil disturbance.

In most regions of the United States no-tillage practices
are becoming increasingly utilized (Phillips 1978, Phillips
et al. 1980), but little information has been developed on
the abundance or importance of nematodes, plant parasitic or
free-living, in no-tillage agroecosystems. Caveness (1974)
reported higher populations of Pratylenchus spp. in tilled
versus no-tillage soils and also in corn roots, but higher
populations of Meloidogyne incognita juveniles and
Helicotylenchus pseudorobustus in no-tillage soils. Thomas
(1978) compared seven tillage regimes and found nematode
densities generally higher in no-tillage plots than in any
of the other tillage practices. Aside from these references,
there is little information specifically on nematodes in no-
tillage systems. However, responses of other soil fauna to
no-tillage practices have been documented. The overall
abundance of soil microarthropods generally is greater in
reduced tillage systems (Weems 1980). In Great Britain,
Edwards (1977) and Edwards and Lofty (1969, 1972) reported
that populations of soil microarthropods in agronomic systems
were severely reduced by conventional plowing but not much
affected by direct drilling (i.e., no-tillage). Price and
Benham (1977) found that cultivation had a greater impact on
surface-dwelling, hemiedaphic arthropods than upon deep soil,
edaphic species in the San Joaquin Valley, California.
Aritajat et al. (1977) demonstrated reductions in soil arthropod and earthworm densities as a result of soil compaction
caused by repeated tractor passage over experimental plots.
Microbial biomass, on the other hand, was found to be significantly greater in no-tillage than in plowed soil (Lynch and
Panting 1979).

    We report here the first year's results for nematode sampl-
ing in an experimental design comparing nutrient cycling under

conventional tillage and no-tillage practices (Stinner and
Crossley 1980). In this research we are attempting to utilize methods and paradigms of ecosystem ecology in an experimental agronomic system. Conventional tillage, no-tillage
and successional old field plots are under study. Comparisons
include nutrient cycling, structure of plant and animal communities, and ecosystem processes (production, consumption,
decomposition). Our general hypothesis is that no-tillage
practice leads to conservation of nutrients and energy.
Excluding tillage is expected to preserve soil processes
relatively intact, at least in comparison with cultivated
soils, permitting the more nearly natural soil processes
which lead to greater nutrient recycling within the system.
Further, we suggest that biotic communities in untilled soil
should resemble those in the more natural, less disturbed
soils of the old-field system.

For nematodes in particular, we predicted that population
response to no-tillage would differ from response to conventional tillage systems, in view of the variety of abiotic
and biotic factors known to influence these animals. No-
tillage practice differs from conventional tillage in that
(1) physical disturbance is minimized and (2) organic material
is not rapidly incorporated into the soil. Organic residues
remain relatively stratified on the soil surface, with slow
incorporation occurring primarily as the result of soil
animal activities. Stratification of organic residues in the
soil surface can influence temperatures, moisture (Phillips
and Young 1973), soil chemistry (Blevins et al. 1980), root
distribution (Maurya and Lal 1980), soil microbial communities (Doran 1980), and soil animals (Edwards 1980). Rates of
mineralization of organic residue are slower under no-tillage
(Stinner and Crossley 1980). In view of these differences

in structural and functional properties, we expected differences in nematode abundance and community structure between
no-tillage and conventional tillage systems.

We report here results of nematode sampling in conventional and no-tillage grain sorghum followed by winter rye.
Our interest was both in plant parasitic and free-living
forms, the latter because of their presumed role in decomposition and nutrient cycling processes. Results of research
in the larger context of nutrient cycling in these small
agroecosystems have been reported separately (Stinner 1981).

SITE DESCRIPTION

Field research was conducted at the University of Georgia's
Horseshoe Bend facility, a 35-acre study area maintained by
the University's Institute of Ecology and located adjacent to
the Oconee River near the campus. The perimeter of the area
is forested and the center is in old-field vegetation except
for the two areas currently being cropped. The site of the
agroecosystem study is a flood plain with well-drained, sandy
clay loam soils, moderately fertile for the Georgia Piedmont.
The two-acre-site was cultivated in 1966 for a millet crop
(Barrett 1968) and had lain fallow until our experiments
were begun in 1978. The area was flooded by river overflow
following a 10-inch rain in 1968. Part of the area was burned
off in 1970 (Odum et al. 1974). The majority of the site was
in woody vegetation at the initiation of the agroecosystem
research. Beginning in May 1978, woody growth was cleared
from the site and herbaceous vegetation was rotary mowed
before the area was subdivided into plots.

METHODS

In May 1978, the two-acre site was subdivided into eight 0.1
ha (quarter-acre) plots. Four plots were randomly assigned
for conventional tillage (CT) and the other four were re-
tained for no-tillage (NT). Each CT plot was disc plowed
to a depth of 8-10 cm and smoothed with a drag harrow (a
procedure repeated in May of 1979). Fertilization and herbi-
cide schedules utilized were those commonly recommended for
the Georgia Piedmont. Fertilizer (6-12-12 NPK) was applied
at a rate of 787 kg · ha^{-1} at the time of planting (early
June in both years). The herbicides Roundup$^{®}$ and Attrex$^{®}$
were applied as pre-emergence weed control (2.2 kg · ha^{-1}
of active ingredient atrazine) each year. No insecticides
were used at any time in these experiments. In June of
each year, grain sorghum (Funk's 522 hybrid) was planted at
a rate of 12 kg seed per ha, in 75-cm rows, using a two-row
fluted coulter no-tillage planter throughout the two-acre
area. Thinning by hand was necessary in late June of the
first year, 1978.

In late November 1978, following harvest, sorghum residue
was disked into the CT plots but left on the surface of the
NT plots. Each plot was then fertilized with 366 kg · ha^{-1}
of 5-10-5 (NPK) fertilizer, and grain rye was drilled into
each plot as a cover treatment. In the spring, the rye was
disk-plowed into the soil in CT plots, but mowed and left on
the surface in the NT plots.

Sampling for nematodes was initiated in July 1979 and was
continued through May 1980, when soybeans were planted in the
plots. Soil samples (2 cm dia X 10 cm deep) were collected
at approximately monthly intervals. Ten randomly located cores
were taken from each of the eight plots at each sampling time.
The cores were taken between the rows at a distance of 25 cm

from the sorghum plants. Nematodes were extracted by the centrifuge-flotation techniques of the University of Georgia's Plant Pathology Extension Service Laboratory.

RESULTS AND DISCUSSION

Densities of all types of nematodes in plots ranged from about 5×10^5 to 18×10^5 per m^2 (Fig. 1). Greatest densities (1.8×10^6 per m^2) were found in the winter months. Nematode densities declined throughout the summer, and lowest densities occurred at the end of November immediately before harvest. Somewhat unexpectedly, total numbers of nematodes were not significantly different ($P > 0.05$) between CT and NT at any of the monthly sampling periods. Figure 2 shows the

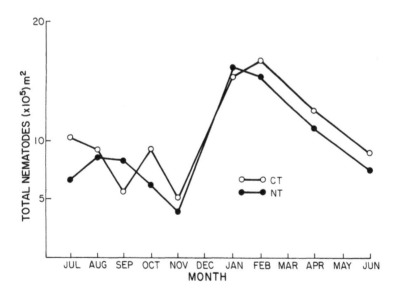

Fig. 1. Total numbers of nematodes/m^2 estimated for agricultural plots under conventional tillage (CT) and no-tillage (NT) cultivation. Results cover the period July 1978–July 1979. Points are means of 10 samples.

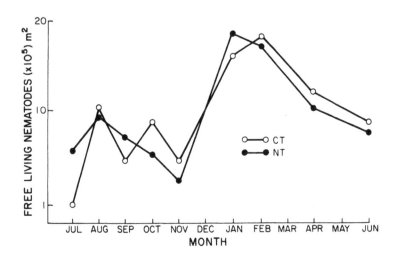

Fig. 2. Numbers of free-living nematodes per m^2 estimated
for agricultural plots under conventional tillage (CT) and
no-tillage (NT) cultivation. Results cover the period July
1978-July 1979. Points are means of 10 samples.

densities of free-living nematodes in CT and NT plots throughout the year. Free-living nematodes comprised roughly 90
percent of the fauna; thus, the seasonal changes in densities
of free-living nematodes followed almost exactly the pattern
of seasonal change for total nematodes (Fig. 1). Total numbers of plant parasitic nematodes (Fig. 3) ranged from 1 X 10^5
to 4 X 10^5 per m^2 during the year. The major differences
between CT and NT practices appeared in plant parasitic nematodes. During the summer, populations were higher in NT plots,
but in winter, CT plots contained higher densities of plant
parasitic forms (Fig. 3). The principal genera of plant parasitic nematodes found in CT and NT plots are listed in Table 1.

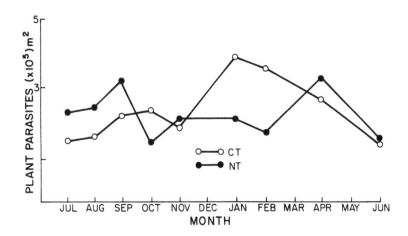

Fig. 3. Numbers of plant parasitic nematodes per m^2 estimated for agricultural plots under conventional tillage (CT) and no-tillage (NT) cultivation. Results cover the period July 1978-July 1979. Points are means of 10 samples.

TABLE 1. Principal Genera of Plant Parasitic Nematodes Found in Conventional and No-tillage Agroecosystem Plots, Horseshoe Bend Research Site, Athens, Georgia, 1978-1979.

Nematode Type	Genera
Spiral	Heliotylenchus, Rotylenchus
Dagger	Xiphinema
Ring	Criconemoides
Lesion	Pratylenchus
Sting	Belonolaimus
Stubby-root	Trichodorus
Root knot	Meloidogyne

The Shannon diversity index H' (Pielou 1966) was calculated for morpho-species of plant parasitic nematodes in the
monthly samples (Fig. 4). Similar trends emerged in both CT
and NT plots. Although initially low, diversity increased
to a maximum in autumn, remained high during winter, and
declined during spring. Plant parasite diversity in CT plots
was higher than in NT plots for the first half of the year
(Fig. 4). The increase in diversity preceded increase in
density by about 3 months.

The most unusual feature of our data appears to be the
large proportion of free-living nematodes in both CT and NT
plots. Typically, plant-parasitic nematodes constitute 50
percent or more of the total nematode fauna (Ferris and Ferris

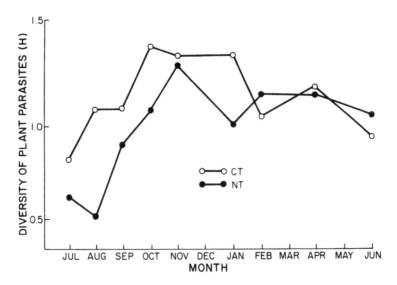

Fig. 4. Shannon diversity index (H') for morpho-species of
plant parasitic nematodes in agricultural plots under conventional tillage (CT) and no-tillage (NT) cultivation,
July 1978-July 1979.

1974). We presume that the large proportion of free-living
forms is due to the lack of cultivation for 12 years preceding our study. Large densities of bacterial and fungal-
feeding nematodes correlate well with the high organic matter
content (Stinner and Hoyt, unpublished data) of the newly
cleared soils. Interestingly, Mononchus, a nematophagous
nematode, was present in most collections, suggesting that
predation might have reduced number of plant parasitic
nematodes.

The sharp increase in nematodes during November and
December can perhaps best be related to two factors: (1)
the input if crop residue in both systems following harvest
of the sorghum and, 2) more favorable soil temperatures (15-
$20°C$) and increased moisture. Summer soil temperatures
typically reach about $40°C$, which quickly dries the soil and
slows nematode activity. The decline of nematode numbers
during spring roughly followed the loss of crop residue
through decomposition (Stinner, in preparation), again related to the large proportion of the fauna being free-living
rather than plant-parasitic.

Numbers of plant parasitic nematodes fluctuated in CT and
NT plots, and were not more abundant in one or the other. The
effects of plowing and mixing the soil appear to be seasonal,
at least during the initial years of cultivation. Little can
be concluded about diversity at this time, except that the
parasite trophic structure appeared to become more complex
in the fall.

Our CT and NT systems represent recently cropped soils and
therefore extrapolation to continuously cultivated systems
is questionable. Yet, the plots in the Horseshoe Bend study
were not atypical for the Georgia Piedmont, in that presently
much reforested land is being put back into cultivation. Thus

studies on newly cleared land are relevant for this region.
Brodie et al. (1970), working in south Georgia, found that
populations of plant parasitic nematodes were initially small
and that 3-5 years were required for damaging levels of feeding to occur. Therefore, we predict that, given several
years, plant parasites will come to comprise a larger proportion of the total nematode fauna.

As we continue to monitor nematodes at Horseshoe Bend, our
focus will be on how CT and NT treatments affect the nematode
community over a 3-4 year period following clearing. No-
tillage produces a stratified upper soil profile compared to
CT, and so we have more recently been sampling each layer
(0-3, 3-4, 10-20 cm) in order to better document nematode
distribution in these two systems. In addition, we need to
concentrate on documenting changes in specific taxa and, in
turn, how these dynamics affect other soil fauna. Finally,
we suggest that the simultaneous measurements of other soil
faunal populations, as well as measurements of the major
processes of primary production, decomposition and consumption,
will provide a broader context in which to interpret nematode
population dynamics.

ACKNOWLEDGMENTS
We gratefully acknowledge the advice and comments of a number
of agronomists, ecologists, horticulturists, nematologists
and plant pathologists in this study but we retain title to
the inadequacies. We thank the University of Georgia's
Institute of Ecology (especially Drs. Eugene P. Odum and
Robert L. Todd) and Division of Entomology (especially Drs.
T. Don Canerday and Preston E. Hunter) for financial and
moral support during the study. We particularly thank Dr.
Richard Snider for critically reviewing the manuscript.

Analyses and symposium travel were supported by a grant (DEB-8007543) from the NSF to the University of Georgia Research Foundation.

LITERATURE CITED

Anderson, R. V., E. T. Elliott, J. F. McClellan, D. C. Coleman, C. V. Cole, and H. W. Hunt. 1978. Trophic interactions in soils as they affect energy and nutrient dynamics. III. Biotic interactions of bacteria, amoebae, and nematodes. Microbial Ecology 4:361-371.

Aritajat, W., D. S. Madge, and P. T. Gooderham. 1977. The effects of compaction of agricultural soils on soil fauna. I. Field investigations. Pedobiologia 17:262-282.

Barrett, G. W. 1968. The effects of an acute insecticide stress on a semi-enclosed grassland ecosystem. Ecology 49:1019-1035.

Blevins, R. L., W. W. Frye, and M. J. Blitzer. 1980. Conservation of energy in no-tillage systems by management of nitrogen. Pages 14-20 in Proceedings of the third annual no-tillage conference. R. N. Gallaher, editor. University of Florida, Gainesville, Florida, USA.

Brodie, B. B., J. M. Good, and C. A. Jaworski. 1970. Population dynamics of plant nematodes in cultivated soil: Effect of summer cover crops in old agricultural land. Journal of Nematology 2:147-151.

Caveness, F. E. 1974. Plant-parasitic nematode population differences under no-tillage and tillage soil regimes in western Nigeria. Journal of Nematology 6:138.

Coleman, D. C., C. V. Cole, R. V. Anderson, M. Blaha, M. R. Champion, M. Clarholm, E. T. Elliott, H. W. Hunt, B. Schaefer, and J. Sinclair. 1977. Analysis of rhizosphere-saprophage interactions in terrestrial ecosystems. Pages

299-309 in Soil organisms as components of ecosystems. U.
Lohm and T. Persson, editors. Ecological Bulletin
(Stockholm) 25:299-309.

Doran, J. W. 1980. Soil microbial and biochemical changes
associated with reduced tillage. Soil Science Society of
America Journal 44:765-771.

Edwards, C. A., and J. R. Lofty. 1969. The influence of
agricultural practices on soil microarthropod populations.
Pages 237-247 in J. G. Sheals, editor. The soil ecosystem.
Systematics Association, No. 8, London, England.

Edwards, C. A. 1980. Interactions between agricultural
practice and earthworms. Pages 3-12 in Daniel L. Dindal,
editor. Soil biology as related to land use practices.
United States Environmental Protection Agency, EPA 560/13-
80-038. Washington, D.C., USA.

Edwards, C. A., and J. R. Lofty. 1972. Effects of insecticides on soil invertebrates. Reports of the Rothamsted
Experiment Station (1971). Harpenden, England.

Ferris, V. R., and J. M. Ferris. 1974. Inter-relationships
between nematode and plant communities in agricultural
ecosystems. Agro-Ecosystems 1:275-299.

Lynch, J. M., and L. M. Panting. 1979. Cultivation and the
soil biomass. Soil Biology and Biochemistry 12:29-33.

Mashkoor, A. M., S. A. Siddique, and A. M. Kahn. 1977.
Mechanism of control of plant parasitic nematodes as a
result of the application of organic amendments to the
soil. III. Role of phenols and amino acids in host
roots. Indian Journal of Nematology 5:27-31.

Maurya, R. P., and R. Lal. 1980. Effects of no-tillage and
ploughing on roots of maize and leguminous crops. Experimental Agriculture 16:185-193.

Odum, E. P., S. E. Pomeroy, J. C. Dickinson, and K. Hutchinson. 1974. The effects of late winter burn on the composition, productivity and diversity of a 4-year-old fallow field in Georgia. Proceedings of the Annual Tall Timbers Fire Ecology Conference 12:399-414.

Phillips, S. H. 1978. No-tillage past and present. Pages 1-5 in Proceedings of the first annual southeastern no-tillage systems conference. J. T. Touchton and D. G. Cummins, editors. Georgia Experiment Station, Athens, Georgia, USA.

Phillips, S. H., and H. M. Young, Jr. 1973. No-tillage farming. Reiman Associates, Inc., Milwaukee, Wisconsin, USA.

Phillips, R. E., R. L. Blevins, G. W. Thomas, W. W. Frye, and S. H. Phillips. 1980. No tillage agriculture. Science 208:1108-1113.

Pielou, E. C. 1966. The measurement of diversity in different types of biological collections. Journal of Theoretical Biology 13:131-144.

Price, D. W., and G. S. Benham. 1977. Vertical distribution of soil inhabiting microarthropods in an agricultural habitat in California, USA. Environmental Entomology 6:575-580.

Sohlenius, B. 1980. Abundance, biomass and contribution to energy flow by soil nematodes in terrestrial ecosystems. Oikos 34:186-194.

Stinner, B. R. 1981. Nutrient budgets and internal cycling of N, P, K, Ca and Mg in conventional tillage, no-tillage and old field systems on the Georgia Piedmont. Ph.D. dissertation. University of Georgia, Athens, Georgia, USA.

Stinner, B. R., and D. A. Crossley, Jr. 1980. Comparison
of mineral element cycling under till and no-till practices: An experimental approach to agroecosystems analysis.
Pages 280-288 in D. L. Dindal, editor. Soil biology as
related to land use practices. United States Environmental
Protection Agency, EPA-560/13-80-038. Washington, D.C., USA.

Thomas, S. H. 1978. Population densities of nematodes under
seven tillage regimes. Journal of Nematology 10:24-27.

Wasilewska, L., E. Paplinska, and J. Zielenski. 1981. The
role of nematodes in decomposition of plant material in a
rye field. Pedobiologia 21:182-191.

Weem, D. C. 1980. The effects of no-till farming on the
abundance and diversity of soil microarthropods. M.S.
thesis. University of Georgia, Athens, Georgia, USA.

Yeates, G. W. 1979. Soil nematodes in terrestrial ecosystems. Journal of Nematology 11:213-229.

INTERACTIONS BETWEEN PHYTOPHAGOUS NEMATODES

Larry Duncan and Howard Ferris

Interactions between nematode species occur at all consumer
trophic levels and in most conceivable habitats. Consequently,
interest in these interactions ranges over a variety of discip-
lines, including parasitology (Hominick and Davey 1972, 1973),
ecology and soil microbiology, (Anderson and Ramsey 1979, Cox
and Coleman 1979, Kozlowska and Domurat 1971), as well as
nematology (Acosta and Ayala 1976, Gay and Bird 1973, Johnson
and Nusbaum 1970). The influences of predatory and competitive
interactions on nematode population dynamics have been investi-
gated and observations may have application to such soil community processes as nutrient cycling and energy transformation
(Anderson et al. 1978).

Sohlenius (1973) has suggested that rhabditid nematodes act
as regulators of bacterial populations during decomposition.
In an attempt to study competition between two supposedly
microbivorous nematodes, Mesodiplogaster sp. and Rhabditis
terricola, he found that Mesodiplogaster sp. actually preyed
upon the concomitant species and may regulate bacterial grazing
(Sohlenius 1968). Competition between species of the genus
Neoaplectana has been used as a criterion for species identification based on the principle of competitive exclusion
(Stanuszek 1972).

Where parasitic nematodes are involved, influence of these
types of interactions on host dynamics has also been examined.
For instance, phytophagous nematodes are attacked by predaceous
members of the phylum, resulting in natural biological control
and indirect influence on primary production (Mankau 1980).
Less attention has been given to competitive relationships

between species of phytophagous nematodes.   Recent attempts at
forecasting nematode damage to crop plants (Ferris et al.
1981, Seinhorst and Kozlowska 1977), emphasize the importance
of understanding how such interactions influence plant growth.
Mathematical models of competitive relationships would be
applicable not only to agricultural pest management systems,
but to ecosystem studies in general.

PREDICTING EFFECTS OF NEMATODE PARASITISM ON PLANT GROWTH
The relationship between plant growth and preplant populations
of phytophagous nematodes is described by a sigmoid shaped
curve for many annual crops (Fig. 1).   A tolerance limit (T)
exists when no reduction in plant growth or yield occurs below

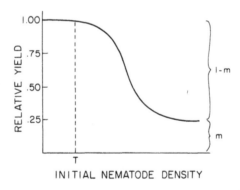

Fig. 1.   General relationship in annual crops between plant
growth and preplant nematode populations.

a certain initial nematode population (P_i). Similarly, there
may be a minimum yield (m) below which no loss occurs regardless of P_i. Characteristics of this curve have been discussed
by Seinhorst (1965, 1972), Oostenbrink (1966), Wallace (1973),
Jones et al. (1978), and others. Yield functions of this form
are useful for predicting crop damage in nematode pest management programs. Data in the range between the tolerance limit
and the nematode population causing minimum yield may be
plotted against logarithmic or square root transformations of
P_i + 1 and predictions made from linear regression models of
these transformed data (Oostenbrink 1966, Seinhorst 1972).

An alternative approach is a biologically descriptive
mathematical model of the functional relationship between
plant growth and P_i (Seinhorst 1965):

$$y = m + (1-m)z^{P-T} \quad \text{for } P > T$$

and

$$y = 1.0 \quad\quad\quad \text{for } P \leq T.$$

According to the model, relative yield (y), the yield at a
given P_i divided by the yield in the absence of nematodes, is
equal to the minimum yield (m) plus some amount of the remaining proportion, 1-m (Fig. 1). The amount of 1-m realized as
yield depends on the term z^{P-T}, in which z represents the proportion of the root system which is not damaged when parasitized
by one nematode, a number slightly less than unity. This parameter is raised to the power which results from subtracting
the tolerance limit from P_i so only those nematodes which
describe the slope of the curve are considered. Seinhorst
used Nicholson's (1933) and Bailey's concept of the competition
curve to derive z^P as an appropriate description of the slope

of yield curves. The competition curve as applied to phytophagous nematodes is based on the principle of overlapping
damage by the parasites (Table 1). If one nematode parasite
causes d proportion of damage to a root, then the proportion
of undamaged root (z) is the quantity 1-d. A second nematode
has only the proportion 1-d available to damage so that the
damage caused by both nematodes is d + d(1-d). The value of
d(1-d) is slightly less than d because it is possible that
some of the tissue encountered by the second nematode was
previously damaged by the first. Subtracting d + d(1-d) from
1.0 leaves $(1-d)^2$ proportion of root undamaged by two nematodes or, in the general case, z^P. As P increases, the value
of z^P decreases so that $(1-m)z^P$ is reduced and yield decreases.

TABLE 1. The relationship between number of nematodes and
proportion undamaged root tissue.

Number Nematodes	Proportion Root Damaged	Proportion Root Not Damaged
1	d	$(1-d) = z$
2	d+d(1-d)	$(1-d)^2$
3	$d+d(1-d)+d(1-d)^2$	$(1-d)^3$
P		$(1-d)^P = z^P$

For certain purposes Seinhorst's model may be a more powerful
pest management tool than the linear model. Because the model
parameters are descriptive not only of empirical observations
(T and m), but also of hypothetical biological processes (z^P),
model validation supports our level of understanding of the
host-parasite relationship. Coefficients derived with linear

regression models, on the other hand, have no comparable biological meaning. As these relatively simple nematode models
are interfaced with other plant growth (or damage) models, it
will be important to describe the biology underlying the parameters to insure that appropriate linkages are developed.
Interactions between the submodels may then be anticipated in
constructing larger, more comprehensive crop models.

Seinhorst (1965) validated the nematode-plant growth model
with published data from field and greenhouse experiments.
Recently, Ferris et al. (1981) developed a computer algorithm
to provide best fit estimates for the T, z, and m parameters
based on least sums of squared residuals. Coefficients of
determination, provided by the algorithm for both field and
greenhouse data, compare favorably with those obtained using
linear regression.

MULTIPLICATIVE HOST DAMAGE RELATIONSHIPS IN MULTISPECIES
INFESTATIONS.

Because monospecific communities of phytophagous nematodes are
rarely observed in nature, a useful expansion of Seinhorst's
model would consider effects of multispecies parasitism on
plant growth. The simplest case would involve closely related
species with similar niche requirements in the same host
tissues, and for which the competition curve represents their
capacity to cause root damage (Table 2). In this special case,
the letters a and b represent different proportions of root
damage and the relationship between respective z^P values for
the two species is multiplicative. In the case of 2 species
with non-identical niche requirements, the relationship may
sometimes be multiplicative for reasons other than overlapping
tissue damage. If, for instance, a vascular pathogen such as
Meloidogyne incognita caused reduction in vascular transport

TABLE 2.　The relationship between number of nematodes of two species and proportion undamaged root tissue.

Number Nematodes		Proportion	Proportion Root
Species A	Species B	Root Damaged	Not Damaged
1	0	a	$(1-a)$
2	0	$a+a(1-a)$	$(1-a)^2$
2	1	$a+a(1-a)+b(1-a)^2$	$(1-a)^2(1-b)$
2	2	$a+a(1-a)+b(1-a)^2+b((1-a)^2(1-b))$	$(1-a)^2(1-b)^2$
P_a	P_b		$z_a^{(P_a)} z_b^{(P_b)}$

while a parasite of the root epidermis such as Tylenchorhynchus
clarus impaired nutrient uptake, the result could be multiplicative in terms of nutrient elements actually transported
to the shoot. In such cases interactions should be considered
in functional rather than physical terms. Because the root is
a functional system, damage to different tissues may exhibit
this relationship, provided the nematode species behave independently.

The majority of studies on nematode interactions with nematodes and other plant pathogens, which report plant growth
data, indicate reasonable agreement with a multiplicative
damage model. The relationship between predicted and actual
yields from 19 such studies (Fig. 2a) were obtained by multiplying relative yields observed in single species inoculations
(i.e., $y = y_1 \times y_2 \ldots \times y_n$) and plotting these values against
actual relative yields observed in multispecies infections.
Multispecies P_i levels of each species were the same as P_i
levels in single species treatments. Deviations of points
from a 45° line represent model and experimental error. The
multiplicative model tends to overpredict damage since the
majority of points lie above the predicted line ($r^2 = 0.70$).
This bias would be expected if no provision is made for the
existence of minimum yield in multispecies infestations.
Because several authors have interpreted such results as evidence of an additive damage relationship, (Bookbinder and
Bloom 1980, Sikora et al. 1972) comparisons were made between
actual yields and yields predicted by adding the damage caused
by single species infestations (i.e., $y = 1 - [(1-y_1)+(1-y_2) \ldots +(1-y_n)]$ (Fig. 2b). Although a strong direct relationship
exists, the additive model explains very little of the vari-
ation of observed from predicted values ($r^2 = .004$). Obser-
vation of apparent additive affects may occur when total plant

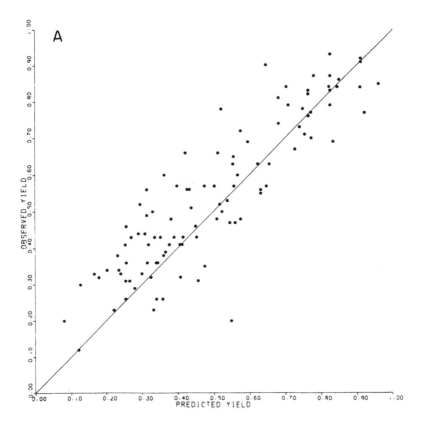

Fig. 2. Relationship between observed relative plant yields
in multiple-pest species infections, and yields predicted from
multiplicative A) or additive B) damage models. 45° line re-
presents exact fit of observed to predicted values. After
Acosta and Ayala (1976), Bookbinder and Bloom (1980), Carter
(1975), Chapman (1959), Goswami and Agarwal (1978), Hague and
Mukhopadhyaya (1979), Heald and Heilman (1971), Johnson (1970),
Johnson and Littrell (1970), Johnson and Nusbaum (1970),
Jorgensen (1970), Kraus-Schmidt and Lewis (1981), Mayol (1970),
Naqvi et al. (1977), Pinochet et al. (1976), Ross (1964),
Sikora et al. (1972), Singh (1976), and Walker and Wallace
(1975).

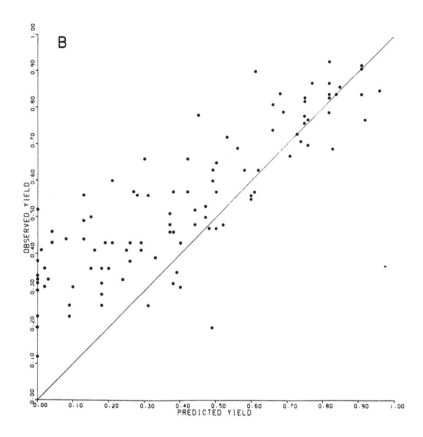

damage is less than 50%, and predictions based on this model
are similar to those of a multiplicative model (Figs. 2a, 2b).
At higher pest densities the predictions diverge and an additive model tends erroneously toward predictions of even greater
loss than does a multiplicative model.

Estimation of the tolerance limit in the multispecies case
may be facilitated by estimation of the relative pathogenicity
of each species provided by the z parameter. If z^T reflects
the minimum of undamaged root system capable of supporting
maximum plant yield, due to excess root production or regen-
erative ability, then this proportion is fixed for a given
type of root tissue and may be termed compensation level (C).
The value of C must be constant for nematode species which
damage identical host tissues under the same environmental
conditions. Similarly, the tolerance capacity, $c = z^{-T}$
(Seinhorst 1965; $y = m + (1-m)cz^P$), is the same for both
species. In practice, due to experimental error, respective
C and c values of individual species will not be identical,
so that average values should be used. The compensation level
for a two species interaction occurs at the level of host
damage, $z_1^{T'} z_2^{T'}$, where T' is a common T value. An estimated
compensation level ($C' = (z_1^{T_1} + z_2^{T_2})/2$) may be used to
derive T'

$$C' = z_1^{T'} z_2^{T'}$$

so that

$$T' = \log C'/(\log z_1 + \log z_2)$$

or in the case of c

$$c' = (c_1 + c_2)/2$$

in which c' represents a general host tolerance capacity.
These methods may be used for any number of concomitant species
providing they damage the same host tissues, and that Seinhorst
model parameter values are known for each species.

Duncan and Ferris (1981, in press) tested a multiplicative yield
model in the simplest case where m = 0.0. Because nematode
damage seldom exhibits this characteristic, two steel needles
of different diameter were used to mechanically damage radish
root systems. Seinhorst model parameters for each needle were
obtained by stabbing seedling root systems with either needle,
and relating plant dry weight at four weeks to the number of
stabs (Ferris et al. 1981). Additional treatments included
root systems stabbed various numbers of times with both needles
to simulate multispecies plant damage (Table 3).

TABLE 3. Predicted and observed[*] relative dry weights of
radish plants subjected to root damage by two different
diameter steel needles (r^2 model predictions = 0.95).

Number Stabs		Relative Yield	
Large Probe	Small Probe	Actual	Predicted
5	5	0.98	0.96
10	5	0.92	0.93
15	5	0.87	0.89
20	5	0.89	0.86
30	5	0.77	0.80
5	15	0.96	0.96
10	15	0.89	0.90
15	15	0.91	0.87
20	15	0.83	0.84
30	15	0.81	0.78
5	30	0.89	0.89
10	30	0.85	0.86
15	30	0.81	0.83
20	30	0.79	0.80
30	30	0.72	0.75
50	50	0.58	0.61

* Treatment data are means of ten replications

The model

$$y = c' \, z_1^{P_1} z_2^{P_2} \qquad\qquad \text{for } y \leq 1.0$$

and

$$y = 1.0 \qquad\qquad \text{for } y > 1.0$$

in which subscripts refer to different needles, was used to
predict relative plant dry weights in the multi-needle treatments. Model predictions (Table 3) accounted for 95% of the
variation between treatment means. The close fit between
observed and predicted plant weights suggest that in the
absence of minimum yield effects, a multiplicative model of
multispecies damage relationships may adequately reflect the
biology of some systems.

THE INFLUENCE OF INTERSPECIES INTERACTIONS ON HOST DAMAGE RELATIONSHIPS

As in the case of tolerance level, a conceptual basis for
minimum yield has not been experimentally tested. It has been
suggested that m results from irregular nematode distribution
in soil so that individual roots escape maximum parasitism
(Seinhorst 1979). Factors influencing the host carrying capacity may also be reflected by m. Evolution of parasitism
toward host–parasite coexistence may involve reduced parasite
reproduction in response to host stress. Controlling signals
might include nutrient availability, available physical root
space or more subtle cues. A number of parthenogenetic species
in the Heteroderidae respond to overcrowding by increasing the
number of adult males which subsequently exit the root tissues
without feeding. The result of such processes is that, either
due to irregular distribution or population responses to host
cues or both, a certain minimum level of host growth may be
anticipated.

For practical purposes, in lieu of experimental evidence
for the basis of minimum yield, hypothetical host-parasite and
parasite-parasite relationships may provide useful estimates
of m in the multispecies case (m'). The parameter m' may be
developed to reflect interspecies competition since concomitant
parasitism influences the population development of the species
involved (Jones et al. 1978). Most frequently, the result is
lower respective final populations, presumably due to decreased
host carrying capacity for each species. Therefore, at given
P_i levels, each species in a concomitant infection may cause
less root damage than in the absence of competing species,
although total damage may be greater in the concomitant
situation.

For instance, it may be reasonable to assume that in many
systems m' lies between values of m for the most and least
pathogenic species, and is influenced by the relative numbers
and virulence of each species. Discounting synergism, it is
unlikely that addition of less virulent nematodes to a system
in which yield relative to a virulent species is at m, will
cause further yield reduction, particularly if each species
affects similar host tissues. Although synergistic interactions
have been demonstrated (Griffin 1980) between Ditylenchus
dipsaci a stem parasite, and M. hapla, a root vascular pathogen,
such observations are infrequent and most likely represent the
exception in nematode disease complexes. Similarly, it is unlikely that addition of a more virulent species will raise the
m value of a specific system.

In the simplest case of two species parasitizing a host, m'
might be estimated by adding m of one species to a proportion
of the difference m_2-m_1. This proportion is related to the
relative capability of the second species compared to the first
either to compete for space in the root or to cause host damage

and influence nutrients available for a second generation of parasites. Factorial experiments designed to demonstrate the competitive capabilities of each species are necessary to elucidate these relationships and will most likely require experiments using a variety of environmental variables, including soil types, moisture regimes and host plants. The number of combinations of nematode species and environmental factors is large enough that estimates of interactions based on relationships between many of these variables is the most realistic approach to the problem. A simplistic approach is to assume an allometric relationship between plant growth and tissues or nutrients available to competing parasites (Seinhorst 1970). Jones et al. (1978) simulated the effects of competition on potato root mass available to <u>Globodera rostochiensis</u> and <u>G. pallida</u> by incorporating the relationship $cz_1^{P_1} z_2^{P_2}$ in a population model. In this case, minimum yield was assumed to be 0.0 and not included in the model.

For nematodes with more than one generation per growing season, the quantity $1-y$ ($y = m + (1-m)z^{P-T}$) for the appropriate P_i of each species may be an adequate reflection of the nematode's ability to influence nutrient availability to its competitor. In such an approach, $(1-y_2)/[(1-y_2) + (1-y_1)]$ might be used as a coefficient of the term m_2-m_1. If damage caused by the second species is sufficiently small compared to the first, the coefficient approaches 0.0 and $m' = m_1 + (m_2-m_1) \times 0.0 = m_1$. Conversely, if the second species is sufficiently numerous and virulent compared to its competitor, the coefficient will approach 1.0 and m' will approach m_2. If initial competition for root space is determined to be an important factor, the term might be further weighted by initial populations as $P_{12}(1-y_2)/[P_{12}(1-y_2)+P_{11}(1-y_1)]$.

The validity of m' as an approximating variable for inter-species competition in a multiplicative yield model was tested by Duncan and Ferris (in press) using cowpea, Vigna sinensis, and the nematode species Meloidogyne javanica and M. incognita. The model used was

$$y = m' + (1-m') \; c' \; z_1^{P_1} z_2^{P_2} \qquad \text{for } y \leq 1.0$$

and

$$y = 1.0 \qquad \text{for } y > 1.0$$

where

$$m' = m_1 + (m_2 - m_1)[(1-y_2)/((1-y_2) + (1-y_1))]$$

and

$$c' = (c_1 + c_2)/2$$

The cowpea variety (Blackeye no. 5) was chosen for its relative resistance to M. incognita, while being fully sus-ceptible to M. javanica. In this way, the model assumption of identical niche requirements was fulfilled using species which produce different levels of effects. Seinhorst model parameter values (m, z, c) for each species were obtained by infesting soil in which individual plants were grown with 13 P_i levels in a geometric progression, and relating P_i to average seed yield in each treatment (Ferris et al. 1981). These parameter values were used in the multispecies model to generate predicted relative seed yields in concomitant species infestations (Fig. 3). Curves on the x and z axes of Fig. 3 represent damage functions of M. javanica and M. incognita respectively. Tolerance limits (M. incognita = 10 eggs/plant, M. javanica = 20 eggs/plant) were similar for the two species, but minimum yields were 0.62 for M. javanica and 0.86 for M. incognita. Predicted relative yields for multispecies infes-tations are shown on the surface portion of the graph.

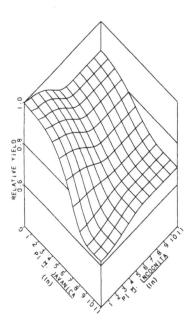

Fig. 3. Predicted relationship between relative dry bean
yield of _Vigna_ _sinensis_ and P_i of _M_. _javanica_ and _M_. _incognita_.

Generally, damage predicted by combinations of the two species
is greater than that of either species alone at its given P_i.
Damage is not strictly multiplicative however because of ad-
justments of population development due to competition. When
P_i of the most damaging species, _M_. _javanica_, approaches m,
predicted yield increases with increasing P_i of _M_. _incognita_
due to competitive effects on the more pathogenic species.
This intriguing possibility has been considered by workers in
both nematology and plant pathology (Jones et al. 1978, Tarr
1972).

Yield predictions tested against observed average yields
of treatments infested with various levels of both species
(Table 4) accounted for 88% of the variation between treatments. Incorporation of the minimum yield parameter (m') in

TABLE 4. Predicted and observed[*] relative bean weights at various P_i levels of two nematode species (r^2 model predictions = 0.88.

Initial Population		Relative Yield	
M. javanica	M. incognita	Actual	Predicted
40	40	1.00	0.96
160	160	0.86	0.85
320	640	0.77	0.77
640	640	0.71	0.73
1280	640	0.70	0.71
2560	640	0.68	0.69
1280	320	0.71	0.71
1280	1280	0.77	0.70
1280	2560	0.67	0.70
20,000	20,000	0.73	0.68

[*]Treatment data are means of ten replications

the model significantly improved model predictions. Predictions based on the assumption of multiplicative damage without inclusion of m' ($y = [m_1 + (1-m_1)z_1^{P_1-T'}] \times [m_2 + (1-m_2)z_2^{P_2-T'}]$) explained only 15% of treatment variation. Lack of fit in this case was primarily due to overestimation of damage at higher P_i levels since the model contains no lower limit (m). By setting m equal to that of the most damaging species (M. javanica) as in the equation, $y = m_1 + (1-m_1)z_1^{P_1-T'}(m_2 + (1-m_2)z_2^{P_2-T'})$, it is possible to account for a greater amount of treatment variation ($r^2 = 0.56$); however, predictions are less accurate than those of the competition model.

Only one treatment (P_i = 20,000 for both species) contained sufficient P_i of M. javanica to test whether M. incognita might show a competitive effect resulting in higher yield. It is interesting that yield in this treatment (0.73) was close to the predicted yield of 0.68 and considerably higher than predicted relative yield for $\underline{M}$. $\underline{javanica}$ alone at P_i = 20,000 (0.62).

CONCLUSION

Acceptable multispecies yield predictions in several experiments were made without prior information about the specific systems in question. Rather, treatment outcomes were forecast from modifications of Seinhorst's biological model, using single species parameter values. The advantage of such an approach is obvious, considering the number of possible nematode-nematode-host combinations to be found in nature. Mathematical simulation of biological interactions occurring in ecosystems allows expansion of the number of independent variables influencing output, without observing each system as a whole. Practical application of such models will most likely optimize the relation between model simplicity and predictive accuracy. For instance, integration of multispecies logistic models of population growth (Volterra 1931, Jones et al. 1978) with Seinhorst's model might provide more realistic and accurate expressions for evaluating multispecies systems. A trade-off would then exist between improved accuracy and complexity, since new parameter values for rates of population increase, equilibrium densities and competition coefficients would need to be empirically obtained. The relationship between plant phenology and nematode damage would also have to be examined in the case of species with more than one annual generation. Such a model, however, could be used to test the validity of simpler models and its usefulness as a research and management tool could be evaluated against the expense of accomodating the added complexity.

LITERATURE CITED

Acosta, N., and A. Ayala. 1976. Effects of Pratylenchus
coffeae and Scutellonema bradys alone and in combination
on guinea yam (Dioscorea rotundata). Journal of
Nematology 8:315-317.

Anderson, R. V., E. T. Elliott, J. F. McClellan, D. C.
Coleman, C. V. Cole, and W. H. Hunt. 1978. Trophic interactions in soils as they affect energy and nutrient dynamics.
III. Biotic interactions of bacteria, amoebae and nematodes. Microbial Ecology 4:361-371.

Anderson, R. V., and K. Ramsey. 1979. Temperature as a niche
dimension decreasing competition in populations of bacteria
feeding nematodes. Journal of Nematology 11:293 .

Bookbinder, M. G., and J. R. Bloom. 1980. Interaction of
Uromyces phaseoli and Meloidogyne incognita on bean.
Journal of Nematology 12:177-182.

Carter, W. W. 1975. Effects of soil texture on the interaction between Rhizoctonia solani and Meloidogyne incognita
on cotton seedlings. Journal of Nematology 7:234-236.

Chapman, R. A. 1959. Development of Pratylenchus penetrans
and Tylenchorhynchus martini on red clover and alfalfa.
Phytopathology 49:357-359.

Cox, D. L., and D. C. Coleman. 1979. Competitive interactions
between bacteria-feeding nematodes. Journal of Nematology
11:296.

Duncan, L. W., and H. Ferris. 1981. Preliminary considerations
of a model of multiple nematode species-plant growth relationships. Journal of Nematology 13:435.

————————————————————————. Validation of a multiple
nematode species-plant damage model. Journal of
Nematology (in press).

Ferris, H., W. D. Turner, and L. W. Duncan. 1981. An algorithm
for fitting Seinhorst curves to the relationship between
plant growth and preplant nematode densities. Journal of
Nematology 13:300-304.

Gay, C. M., and G. W. Bird. 1973. Influence of concomitant
Pratylenchus brachyurus and Meloidogyne spp. on root penetration and population dynamics. Journal of Nematology
5:212-217.

Goswami, B. K., and D. K. Agarwal. 1978. Interrelationships
between species of Fusarium and root-knot nematode,
Meloidogyne incognita in soybean. Nematologia Mediterranea
6:125-128.

Griffin, G. D. 1980. Interrelationship of Meloidogyne hapla
and Ditylenchus dipsaci on resistant and susceptible
alfalfa. Journal of Nematology 12:287-293.

Haque, M. S., and M. C. Mukhopadhyaya. 1979. Pathogenicity
of Macrophomina phaseoli on jute in the presence of
Meloidogyne incognita and Hoplolaimus indicus. Journal
of Nematology 11:318-320.

Heald, C. M., and M. D. Heilman. 1971. Interaction of
Rotylenchulus reniformis, soil salinity, and cotton.
Journal of Nematology 3:179-182.

Hominick, W. M., and K. G. Davey. 1972. The influence of host
stage and sex upon the size and composition of the population
of two species of thelastomatids parasitic in the hindgut
of Periplaneta americana. Canadian Journal of Zoology
50:947-954.

————————————————————————. 1973. Food and the spatial
distribution of adult female pinworms parasitic in the
hindgut of Periplaneta americana L. International Journal
of Parasitology 4:759-771.

Johnson, A. W. 1970. Pathogenicity and interaction of three

nematode species on six bermuda grasses. Journal of
Nematology 2:36-41.

Johnson, A. W., and R. H. Littrell. 1970. Pathogenicity of
Pythium aphanidermatum to Chrysanthemum in combined inoculations with Belonolaimus longicaudatus or Meloidogyne
incognita. Journal of Nematology 2:255-259.

Johnson, A. W., and C. J. Nusbaum. 1970. Interactions between
Meloidogyne incognita, M. hapla and Pratylenchus brachyurus
in tobacco. Journal of Nematology 2:334-340.

Jones, F. G. W., R. A. Kempton, and J. N. Perry. 1978.
Computer simulation and population models for cyst-nematodes.
Nematropica 8:36-56.

Jorgensen, E. C. 1970. Antagonistic interaction of Heterodera
schachtii Schmidt and Fusarium oxysporum (Woll.) on sugarbeets. Journal of Nematology 2:393-398.

Kozlowska, J., and K. Domurat. 1971. Research on the biology
and ecology of Panagrolaimus rigidus (Schneider) Thorne.
Ekologia Polska 19:715-722.

Kraus-Schmidt, H., and S. A. Lewis. 1981. Dynamics of concomitant populations of Hoplolaimus columbus, Scutellonema
brachyurum, and Meloidogyne incognita on cotton. Journal
of Nematology 13:41-48.

Mankau, R. 1980. Biological control of nematode pests by
natural enemies. Annual Review of Phytopathology
18:415-440.

Mayol, P. S., and G. B. Bergenson. 1970. The role of
secondary invaders in Meloidogyne incognita infection.
Journal of Nematology 2:80-83.

Naqvi, Q. A., M. M. Alam, S. K. Saxena, and K. Mahmood. 1977.
Effect of spinach mosaic virus, root-knot and stunt nematodes on growth of sugarbeet. Nematologia Mediterranea
5:145-149.

Nicholson, A. J. 1933. The balance of animal populations.
   Journal of Animal Ecology 2:132-178.

Oostenbrink, M. 1966. Major characteristics of the relation
   between nematodes and plants. Mededelingen Labdbouwhoge-
   school Wageningen 66:1-46.

Pinochet, J., D. J. Raski, and A. C. Goheen. 1976. Effects
   of Pratylenchus vulnus and Xiphinema index singly and combined on vine growth of Vitis vinifera. Journal of
   Nematology 8:330-335.

Ross, J. P. 1964. Interaction of Heterodera glycines and
   Meloidogyne incognita on soybeans. Phytopathology 54:304-
   307.

Seinhorst, J. W. 1965. The relationship between nematode
   density and damage to plants. Nematologica 11:137-154.

----------------. 1970. Dynamics of populations of plant
   parasitic nematodes. Annual Review of Phytopathology
   8:131-156.

----------------. 1972. The relationship between yield and
   square root of nematode density. Nematologica 18:585-590.

----------------. 1979. Nematodes and growth of plants: for-
   malization of the nematode-plant system. Pages 231-256 in
   F. Lamberti and C. E. Taylor, editors. Root-knot nematodes
   (Meloidogyne species): Systematics, biology and control.
   Academic Press, New York, New York, U.S.A

Seinhorst, J. W., and J. Kozlowska. 1977. Damage to carrots
   by Rotylenchus uniformis, with a discussion on the cause of
   increase of tolerance during the development of the plant.
   Nematologica 23:1-23.

Sikora, R. A., D. P. Taylor, R. B. Malek, and D. I. Edwards.
   1972. Interaction of Meloidogyne naasi, Pratylenchus
   penetrans and Tylenchorhynchus agri on creeping bentgrass.
   Journal of Nematology 4:162-165.

Singh, N. D. 1976. Interactions of Meloidogyne incognita
and Rotylenchulus reniformis on soybean. Nematropica
6:76–81.

Sohlenius, B. 1968. Studies of the interactions between
Mesodiplogaster sp. and other rhabditid nematodes and a
protozoan. Pedobiologia 8:340–344.

————————. 1973. Structure and dynamics of populations
of Rhabditis (Nematoda: Rhabditidae) from forest soil.
Pedobiologia 13:368–375.

Stanuszek, S. 1972. Adaptation of studies on interspecies
competition for differentiating nematode species of the
genus Neoaplectana Steiner, 1929 (Rhabditoidea; Steiner-
nematidae). International symposium of nematology (11th)
European Society of Nematologists, Reading, United Kingdom.

Tarr, S. A. J. 1972. Principals of plant pathology.
Winchester Press, New York, New York, U.S.A.

Volterra, V. 1931. Variations and fluctuations of the
number of individuals in animal species living together.
Pages 409–448 in R. N. Chapman. Animal ecology. McGraw-
Hill, New York, New York, U.S.A.

Walker, G. E., and H. R. Wallace. 1975. The influence of
root-knot nematode (Meloidogyne javanica) and tobacco
ringspot virus on the growth and mineral content of
french beans (Phaseolus vulgaris). Nematologica 21:455–462.

Wallace, H. R. 1973. Nematode ecology and plant disease.
Crane, Russak, New York, New York, U.S.A.

Part II
Decomposition

ROLE OF NEMATODES IN DECOMPOSITION

G. W. Yeates and D. C. Coleman

INTRODUCTION

In the past decade there has been considerable interest in
assessing soil nematode populations and determining their
metabolic activity in relation to other soil organisms and to
primary production. Nematodes generally contribute less than
1% of total soil respiration (Yeates 1979), and only a few
percent can be attributed to all soil fauna (Reichle 1977).
As Macfayden (1978) notes, such approaches are of limited
general value because of the complexity and diversity of
specific physiology under field conditions. Decomposition is
neither a simple combustion-like process nor confined to the
soil. In fact, a simple correlation of metabolic activity
(such as CO_2 output) with nutrient cycling is occasionally
misleading, or even erroneous.

In this review we relate soil nematode populations to both
decomposition processes and nutrient cycling in which decomposition and plant growth are complementary. This implies
that at least some soil nematodes have a beneficial effect on
plant growth and relates nematodes to the dynamics of organic
matter in the rhizosphere (Coleman 1976, Sauerbeck and Johnen
1977), as well as to current developments in effects of
faunal grazing on microbial populations (Addison and Parkinson
1978, Hanlon and Anderson 1979). A small grazing component
may play a large part in ecosystem regulation (Lee and Inman
1975). Twinn (1974) commented on the importance of parallel
studies of both nematodes (grazers) and microflora. In this
paper we review both laboratory and field-based studies, in a
variety of habitats, both natural and man-made, to examine

the principal processes which affect the roles of nematodes
in decomposition.

Decomposition studies in an ecosystem context are only now
coming "of age" in several regions of the globe. To adequately study decomposition requires considerable background in
biology, chemistry, and physics, perhaps one of the legacies
as to "modus operandi" from the International Biological
Program (IBP).

NEMATODE FEEDING HABITS AND REPRODUCTION
Nematodes feed on a wide range of foods and a basic trophic
grouping is: bacterial feeders, fungal feeders, plant
feeders, predators and omnivores. Nicholas (1975) has reviewed classification of feeding habits, and Figure 1

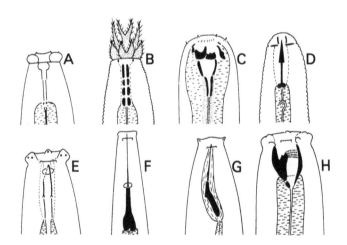

Fig. 1. Head structures of a range of soil nematodes. A)
Rhabditis (bacterial feeding); B) Acrobeles (bacterial feeding); C) Diplogaster (bacterial feeding, predator); D)
tylenchid (plant feeding, fungal feeding, predator); E)
Dorylaimus (feeding poorly known, omnivore); F) Xiphinema
(plant feeding); G) Trichodorus (plant feeding); H) Mononchus
(predator).

illustrates anterior structures associated with various modes
of feeding. The classification used by Paramonov (1968) including eusaprobes (associated with decomposition and not
directly pathogenic), dyssaprobes (feed in decomposing
material, but may enter healthy tissue temporarily),
pararhizobes (found in rhizosphere, may sometimes damage roots),
and phytoparasites (invade plants and feed on their tissues)
has great ecological value. However, the major problem is
that feeding habits of soil nematodes are both complex and
poorly known. Among the Tylenchida, for example, the
Heteroderidae, Hoplolaimidae, Tylenchorhynchidae, and
Criconematidae as well as Anguina and Radinaphelenchus feed
solely on higher plants, but within the Neotylenchidae,
Aphelenchoididae, Ditylenchus and Tylenchus a wide range of
feeding habits are known, including root feeding, root hair
feeding, fungal feeding, and association with insects. Without in vivo observations it is not possible to assess trophic
relations in a particular situation. In addition, feeding
habits may change during the life cycle. From their distribution it appears that many Tylenchida are associated with
fungi rather than plant roots and this gives them an important role in decomposition.

Nematodes prey on other soil animals (e.g., Nygolaimus on
enchytraeids) and in turn are prey for others (e.g., tardigrades). Less well known examples are the isopod (Oniscus
asellus grazing on nematodes in sewage sludge (Brown et al.
1978), the shrimp Crangon feeding on nematodes among sand
grains (Gerlach and Schrage 1969).

Among the Dorylaimida there are a few established plant
feeders (Trichodorus, Longidorus, Xiphinema), predators
(Mononchidae, Nygolaimus) and fungal feeders (some
Tylencholaimus spp.), but most of these relatively large

nematodes have undetermined feeding preferences. Yeates (1973) suggested that they may be autochthonous (slow) decomposers, in contrast to zymogenous (rapid) decomposers (Rhabdita, etc.). McKercher et al. (1979) suggested that Dorylaimidae may play a part in phosphorous transformations greater than their relative biomass suggests.

To feed and reproduce nematodes must have a suitable physical environment. Jones et al. (1969) related the distribution of some plant nematodes to soil structure, Heterodera spp. occurring in fine and coarse textured soils but the larger Trichodorus and Longidorus only in soils with 80% or more of coarse particles. Recent volcanic soils, in which particles are more angular than older soils, have been found to contain a more diverse range of genera, including Acrobeles, than older, more weathered soils (Yeates 1980a). Recent microcosm studies have shown that soil nematode growth is less in a fine textured than in a coarse textured soil with more habitable pores for nematodes; a similar result was found in both Mesodiplogaster + Pseudomonas (MP) and Mesodiplogaster + Pseudomonas + Acanthamoeba (MPA) interactions (Elliot et al. 1980) (Fig. 2). Thus, the trophic interactions between soil fauna in soils of differing texture must be examined to determine the amount of faunal grazing on decomposer populations and the resultant nutrient release.

ECOSYSTEMS AND TOTAL NEMATODES

It is well known that an increase in plant feeding nematode populations leads to a decrease in crop yields (Fig. 3a) (Sykes 1979). However, there may be a positive correlation between total nematode population and primary production. Figure 3b-d gives such results for three grassland studies.

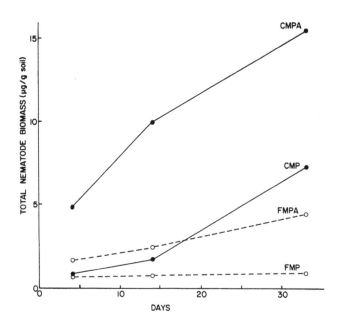

Fig. 2. Influence of soil texture (C, sandy clay; F, clay) on
development of nematode (M, _Mesodiplogaster lheritieri_) biomass in microcosms containing a bacterium (P, _Pseudomonas_
cepacia) with or without an amoeba (A, _Acanthamoeba polyphaga_)
at 26°C (after Elliott et al. 1980).

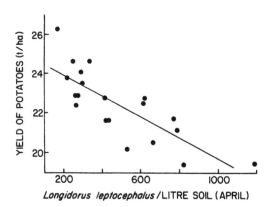

Fig. 3a. Relationship between potato yield and numbers of
Longidorus leptocephalus in soil (r = -0.80,99%) (after Sykes
1979).

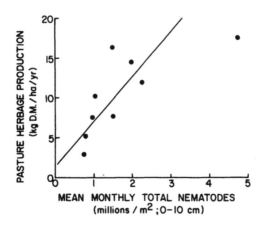

Fig. 3b. Relationship between mean monthly total nematodes and total annual pasture herbage production for nine soils under grazed pasture at least 5 years old (r = +0.71, 95%) (after Yeates 1979).

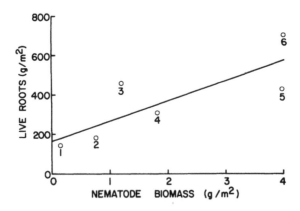

Fig. 3c. Relationship between total nematode biomass and weight of live roots at 6 US/IBP grassland sites in 1972 (r = +0.81, 95%); 1) Richland, 2) Las Cruces, 3) Pasukuska, 4) Nunn, 5) Cottonwood, 6) St. Lanatius (data from French 1979).

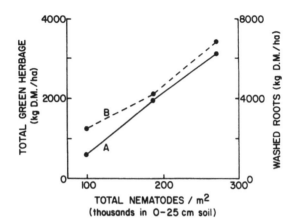

Fig. 3d. Relationship between mean total nematode populations and A) standing green herbage, and B) washed roots, in three treatments of a grazing trial at Armidale, Australia (after King and Hutchinson 1976).

Such a clear correlation is not always found. In tussock grasslands, where phosphorus is probably the limiting plant nutrient, there is a significant correlation between total nematodes and available P (Fig. 3e); at sites with similar soils this can be regarded as a nematode:production correlation. Freckman and Mankau (1979) found a positive correlation between the distribution of nematodes and the root biomass of four desert shrubs.

In forests the situation is less clear. Perhaps the extreme is a significant negative correlation between net primary production and nematode abundance (Kitazawa 1971) (Fig. 3f). In such situations the driving variable for nematodes in 0-20 cm soil is more likely to be some litter function rather than total primary production. A similar effect is seen in the Japanese IBP results in which the coniferous Shygayama area had 6.87 X 10^6 nematodes/m^2 and the evergreen oak Minamata area 3.23 X 10^6/m^2; the litter production at the two sites was

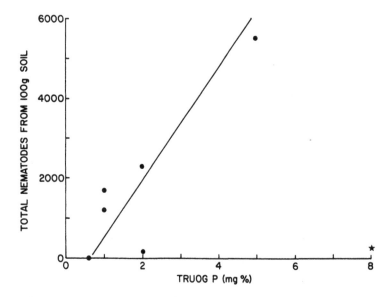

Fig. 3e.   The relationship between total nematode abundance and Truog P in seven soils from a tussock-grassland climosequence; if the gley soil (*) is omitted, r = +0.89 (98%); but if all points are considered, r = +0.20 (from Yeates 1974, Molloy and Blakemore 1974).

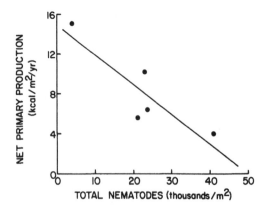

Fig. 3f.   Relationship between net primary production and soil nematode population in five tropical forests (r = -0.88, 95%) (data of Kitazawa 1971).

409 and 732 $g/m^2/yr$ and net aboveground primary production 7.63
and 18.38 t/ha/yr respectively (Kira et al. 1978, Kitazawa
1977). In two Australian subtropical forests, Plowman (1979)
found similar litter falls (912.8 and 886.4 $g/m^2/yr$) in a rain
forest and wet sclerophyll forest but a five-fold difference
in the nematode populations (805.8 and 154.5/m^2 respectively).
In contrast, Huhta and Koskenniemi (1975) found a correlation
between plant production and nematodes (Table 1), but estimated
numbers, biomass, and respiration give different trends.

TABLE 1. Primary Production in Two Finnish Spruce Forests,
with Estimates of Nematode Populations and Activity in the Soil
Organic Horizons of Each (After Huhta and Koskenniemi 1975)

Forest type	Lammi Oxalis-Myrtillus	Oulanka Hylocomium-Myrtillus
Net primary production (excluding roots) ($g/m^2/yr$)	889	420
Litter production ($g/m^2/yr$)	292	143
Total nematodes (thousands/m^2)	1436	1125
Nematode biomass (mg/m^2)	178	200
Nematode respiration (ml $O_2/m^2/yr$)	439	329

In South Carolina, Lane (1975) followed the effects of
forest conversion from hardwoods to pines over seven years;
the best correlation in his data is +0.98 (95%) between winter
total nematode populations and forest litter in the control

plots; in the treated plots the correlation was +0.79. Perhaps the nematode fauna need to reach equilibrium before
sampling. This is the converse of models for economically
important nematodes used by Seinhorst (1970) and Jones and
Kempton (1978) in which gross changes in populations are predicted.

Even nematode populations in noncropped ecosystems differ
markedly from year to year (Banage 1966, Popovici 1977, Huhta
et al. 1979, and Yeates 1980b). In the case of New Zealand
grazed pastures, there was a significant correlation over
36 months between total nematode abundance and aboveground
production, the usual temperature and moisture effects not
being directly significant in the regression. Similar relations between primary production and nematode populations
may underlie the differences found in other studies.

Thus in the ideal situation, nematode ecology would be
studied in a stable ecosystem, if such an entity exists, over
a period of years with aspects of primary production assessed.
Fluctuating ecosystems such as crop rotations give valuable
data on some aspects of population ecology. The major problem in writing this review has been the lack of comprehensive
coordinated site data, such as total production of shoots and
roots, soil carbon and nitrogen levels or long-term sampling.

Given a positive correlation between nematode abundance
and primary production it follows that there should be a
similar relationship between nematode abundance and decomposition, provided a similar combination of trophic groups
exists.

NEMATODES AND DECOMPOSITION SITES
There is a long history of work on nematodes in sites of
decay, ranging from the work of Paesler (1946) and Sachs

(1950) in dung pats to the population estimates of 16 million
nematodes/m^2 in the intertidal region (Teal and Wieser 1966)
and 380 million bacterial and fungal feeders/m^2 in leaf litter
(Wasilewska 1979). Although large populations/m^2 can be calculated, they are usually extrapolations from small foci of
intense decomposer activity. In general, the distribution of
nematodes in the soil profile reflects the distribution of
organic matter produced in the ecosystem--from dung pats and
decomposing litter on the surface to senescent roots a meter
or more down. Grassland may be uniform horizontally or tus-
socks may be clumped; forests have a variety of canopy, shrub,
and herb inputs; crops are generally spaced and in tundra
there is a mosaic of vegetation and bare ground. All these
conditions interact with factors such as other soil organisms
(Fig. 4), depth (e.g., soil horizons; see Gould et al. 1979),
season and year to produce a complex nematode population
structure.

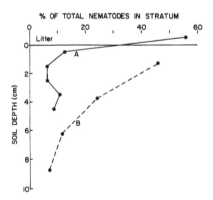

Fig. 4. Effect of litter incorporation on the vertical distribution of nematodes in grassland. A) Pawnee IBP site,
Colorado; no earthworms, litter layer present (from Anderson
1978), B) mean of 9 grazed pastures in New Zealand; all sites
had earthworms but no litter layer (from Yeates 1980a).

CURRENT STATUS OF WORK

Field work has been reviewed in three recent papers (Wasilewska 1979, Yeates 1979, Sohlenius 1980). Currently, papers are appearing in which microcosms are used to manipulate conditions and measure responses to known changes and stresses. Development of such work appears essential in assessing the role of nematodes in decomposition, and it is to such studies that we now wish to mainly address the balance of our paper.

NEMATODE:FOOD INTERACTIONS

As noted in the previous section, populations of certain plant parasitic nematodes are considered to be increasingly deleterious to plant growth, with increments in nematode numbers. As we will endeavor to show, such trends are too simplistic, and a more dynamic view is required. Some examples follow:

Plant Feeding Nematodes

When experimental plant inoculations are made, low soil populations cannot be measured accurately. To remedy this, Jones (1957) set up a pot experiment, raising beet seedlings in a sterilized potting mix. Cysts of Heterodera schactii containing an average of 30 eggs each were added on a log series (1, 5, 25, 125, and 625). Seedlings were harvested after 4 to 5 months. Total yield of roots plus tops was significantly greater with 1 cyst per pot than in the un-inoculated controls or with higher numbers. This sort of stimulation by low-level feeding, which has been reported for many other nematode/plant combinations, may be an important factor for consideration in several "plant-animal interaction" experiments currently in progress.

Fungal Feeding Nematodes

Wasilewska et al. (1975) cultured Alternaria tenuis on agar,
assessing nematodes (Aphelenchus avenae) and mycelia every
few days. There was some inhibition of mycelial growth in
cultures with, versus those without nematodes (Fig. 5).
Behavior of the same fungal grazer (A. avenae) and a fungus
(Rhizoctonia solani) was examined in soil microcosm studies
(Trofymow and Coleman, this symposium). The microcosms with
nematodes feeding respired significantly less than those with
only fungus metabolizing pure cellulose substrate (Fig. 6).
In both cases nematode grazing depressed fungal activity, but
the generality of this effect is unknown.

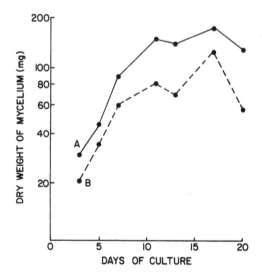

Fig. 5. Development of the fungus Alternaria tenuis at 27°C.
A) A. tenuis alone, B) A. tenuis with Aphelenchus avenae
(after Wasilewska et al. 1975).

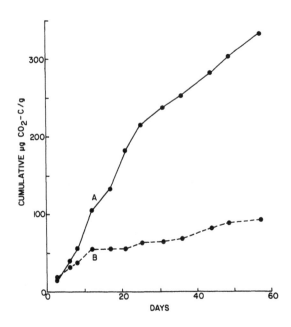

Fig. 6. Effect of nematode (Aphelenchus avenae) grazing on total respiration of soil microcosms at 23°C. A) Rhizoctonia solani alone, B) R. solani with A. avenae (Trofymow and Coleman, this symposium).

Bacterial Feeding Nematodes

Using microcosms, a series of experiments were conducted using a rhizosphere bacterial isolate (Pseudomonas sp.) and an omnivorous nematode Mesodiplogaster lheritieri. In all cases, those microcosms with bacteria and nematodes had greater respiratory activity (CO_2 output) than those with bacteria alone. Bacterial populations were markedly reduced, where grazed upon, with a significant amount of nitrogen and phosphorus remineralized only after 10 days (Anderson et al. 1981).

Using a more enriched substrate (sewage sludge addition to a silt loam), a series of microcosms were inoculated with Pseudomonas fluorescens and some with the bacteria and the bacterial feeding nematode Pelodera punctata (Abrams and

Mitchell 1980). Total respiratory activity was also increased here, in comparison with that of bacteria alone
(Fig. 7). The stimulation was greater than in the mostly
mineral soil microcosms of Coleman et al. (1976). Abrams and
Mitchell (1980) gave no information on the mineralization
(into the soil solution) of inorganic nitrogen and phosphorus
but this would not be likely, given the rapid regrowth of
bacteria in the high-organic matter substrate.

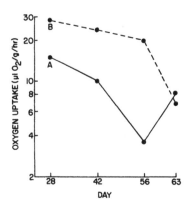

Fig. 7. Oxygen uptake of sewage sludge microcosms at 5° C.
A) Pseudomonas fluorescens alone, B) P. fluorescens with
Pelodera punctata (after Abrams and Mitchell 1980).

Microcosm Studies, Including Plant Growth

To determine possible effects of faunal grazing on microflora
as it influences plant growth, Bååth et al. (1978) investi-
gated growth responses of Scots pine seedlings (Pinus
sylvestris L.) raised in a humus-sand mixture. The pots
were watered with four combinations of nutrient solutions:

(1) complete nutrient solution minus nitrogen; (2) glucose;
(3) nitrogen; and (4) glucose plus nitrogen. Abundance and
biomass of bacteria, fungi, arthropods, and other invertebrates, including nematodes, were determined after 398 d
when pine root and shoot biomass and N concentrations were
also determined. Bååth et al. found considerable stimulation
of bacterial yeast, and fungal growth, but only fungal-grazing
nematodes and a few bacterial feeders increased in any of the
treatments. Essentially no new tree growth could be attributed to faunal grazing activity. However, as the "humus"
material had a relatively low N content (C:N ratio 44:1),
it is likely that nitrogen would be immobilized in such a
system. Blue grama (Bouteloua gracilis) grass seedlings
showed significant increases in shoot N content in grazed
microcosms (Elliott et al. 1979). This occurred only in
those microcosms with bacteria and amoebae and with moderate
to high levels of ammonium fertilizer. With a more favorable
C:N ratio in the soil organic matter, roots may more readily
take up nutrients.

Experiments with laboratory cultures have shown that grazing of Rhabditis oxycerca on Bacillus subtilis reduces the
antagonism of B. subtilis to Fusarium solani and thus allows
greater fungal growth and damage to Pinus nigra seedlings
(Palmisano and Turchetti 1975).

NEMATODE ACTIVITIES IN AN ECOSYSTEM CONTEXT
A series of studies are reviewed below to give information
on a wide range of community/ecosystem responses which indicate the diversity of effects which have been observed in
primary production and decomposition when ecosystems have
been experimentally manipulated.

Nematode Effects on Primary Production

Smolik (1977) applied a systemic nematicide, Vydate® (oxamyl, an oxime carbamate) to a South Dakota mixed-grass prairie. He followed subsequent growth and regrowth on three cuttings from experimental and control plots, during the growing season (Table 2). The increases in dry matter yield were far

TABLE 2. Effect of Vydate Treatment on Nematode Numbers and Herbage Yield of Range Grasses at Cottonwood IBP Site (After Smolik 1977).

Date	Percent reduction in nematodes			% increase in herbage dry weight yield
	Plant feeding	Predaceous	Saprophagous	
6 Jul 72	89	61	58	59
21 Jul 72	89	73	77	28
28 Sep 72	--	--	--	45
5 Apr 73	82	53	52	--

in excess (ca. 10-12X) of the calculated amount of root material ingested (consumed) by nematodes. Recent information on other effects of Vydate indicate that there is probably plant growth stimulation and a fertilizer effect at the concentration used by Smolik (Freckman and Van Gundy, pers. comm.). However, Smolik noted that while there was a reduction of almost 90% of the plant feeding nematodes in 0-10 cm soil 12 months after treatment, about 60% of the saprophagous and predaceous trophic groups were killed as well.

To more fully examine the spectrum of soil biota affected, Stanton et al. (1981) applied a systemic nematicide, Furadan® (carbofuran, an N-methyl carbamate) on plots in the Pawnee site, a Colorado shortgrass prairie containing predominantly

blue grama and fringed sagewort (<u>Artemisia</u> <u>frigida</u>). The
only significantly greater root biomass (140 g or ca. 25%
larger) was in the nematicide treated plots late in the growing
season (September) in both years (1975-6) of the field study.
Significant decreases were noted in most biota except bacteria.
In addition to a mortality of about 90% for total nematodes,
there were major decreases in soil microarthropods and a 50%
decrease in fungal propagules, but not in mycorrhizal infection frequencies. It is apparent from these studies that a
more discriminating faunicide is needed to be certain of
causative factors affecting primary production.

Nematode Roles in Community Food-Chains

Another aspect of crucial importance to the role of nematodes
in systems dynamics is community food-chain or food-web structure. We give a brief synopsis of studies on desert litter
decomposition, as it is covered more fully elsewhere in the
symposium. Using a series of insecticide and fungicide
treatments, Santos et al. (1981) manipulated populations of
microarthropods, fungi, and nematodes. With elimination of
microarthropods (primarily Prostigmata:Tydeidae) there were
increased numbers of bacterial feeding nematodes and reduced
bacterial numbers. With elimination of both nematodes and
microarthropods, bacterial numbers were greater than in untreated controls. Fungal hyphae increased in insecticide-
treated (microarthropod-free) treatments.

By a series of additional manipulations, Santos et al.
(1981) and Whitford et al. (this symposium) concluded that
the 40% reduction in decomposition in mite-excluded plots
(principally Tydeidae) was the result of a removal of predatory pressure on the dominant bacterial grazers, cephalobid
nematodes. Coleman et al. (1978) noted bacterial-grazing

nematode populations can build up to a considerable level
causing net immobilization of nutrients unless continual
predation pressure exists.

A hitherto overlooked area of faunal interactions has been
found by Yeates (1981). Sampling total nematodes in three
soil types under grass, Yeates observed ca. 50% decrease in
total numbers of nematodes in New Zealand pastures with earthworms, compared with those without (Fig. 8). There are
several possible explanations, such as limiting amounts of
food leading to considerable drop in the nematodes. Another
possibility is that earthworms, ingesting large amounts of
soils, or other substrates containing nematodes, simply trit-
urate the nematodes enough to significantly decrease their

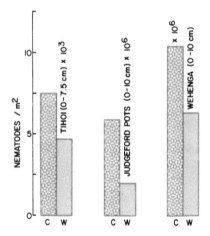

Fig. 8. Effect of addition of earthworms on total nematode
abundance (/m^{-2}, depth as indicated) in three New Zealand
soils; in each case, the right-hand column (W) represents
with earthworm situation (Yeates 1981).

standing crop. For example, Phillips (in Piearce and Phillips 1980) noted that specimens of <u>Lumbricus</u> <u>terrestris</u> that fed on cattle dung containing large numbers of nematodes had active nematodes in their pharyngeal and esophageal regions; but none were found in material present in crops and gizzards or in castings from the earthworm.

CONCLUSION

Results from natural and perturbed field sites, and experimental trophic studies including microcosms show many interactions among nematodes and other soil biota. These interactions often result in enhanced nutrient circulation and eventually higher productivity. From this review it is clearly important to use a range of experimental techniques, in laboratory and in field, rather than relying solely on one level of resolution when investigating the ecological roles of nematodes in the soil.

ACKNOWLEDGMENTS

Support for preparation of this paper was provided by NSF Grant DEB 80-04193 to Colorado State University, and the Department of Scientific and Industrial Research, New Zealand Government.

LITERATURE CITED

Abrams, B. J., and M. J. Mitchell. 1980. Role of nematode-bacterial interactions in heterotrophic systems with emphasis on sewage sludge decomposition. Oikos 35:404-410.
Addison, J. A., and D. Parkinson. 1978. Influence of collembolan feeding activities on soil metabolism at a high arctic site. Oikos 30:529-538.

Anderson, R. V. 1978. Free-living nematode population dynamics: effects on nutrient cycling. Ph.D. Dissertation, Colorado State University, Fort Collins, Colorado, USA.

Anderson, R. V., D. C. Coleman, C. V. Cole, and E. T. Elliott. 1981. Effects of the nematodes Acrobeloides sp. and Mesodiplogaster lheritieri on substrate utilization and nitrogen and phosphorus mineralization in soil. Ecology 62:549-555.

Bååth, E., U. Lohm, B. Lundgren, T. Rosswall, B. Söderström, B. Sohlenius, and A. Wiren. 1978. The effect of nitrogen and carbon supply on the development of soil organism populations and pine seedlings: a microcosm study. Oikos 31:153-163.

Banage, W. B. 1966. Nematode distribution in some British upland moor soils with a note in nematode parasitizing fungi. Journal of Animal Ecology 35:349-361.

Brown, B. A., B. L. Swift, and M. J. Mitchell. 1978. Effect of Oniscus asellus feeding on bacterial and nematode populations of sewage sludge. Oikos 30:90-94.

Coleman, D. C. 1976. A review of root production processes and their influence on soil biota in terrestrial ecosystems. Pages 417-434 in J. M. Anderson and A. Macfayden, editors. The role of terrestrial and aquatic organisms in decomposition processes. Blackwell Science Publications, Oxford, England.

Coleman, D. C., R. V. Anderson, C. V. Cole, E. T. Elliott, L. Woods, and M. K. Campion. 1978. Trophic interactions in soils as they affect energy and nutrient dynamics. IV. Flows of metabolic and biomass carbon. Microbial Ecology 4:373-380.

Elliott, E. T., D. C. Coleman, and C. V. Cole. 1979. The influence of amoebae on the uptake of nitrogen by plants

in gnotobiotic soil. Pages 221-229 in J. L. Harley and
R. S. Russell, editors. The soil-root interface.
Academic Press, London, England.

Elliott, E. T., R. V. Anderson, D. C. Coleman, and C. V. Cole.
1980. Habitable pore space and microbial trophic interactions. Oikos 35:327-335.

Freckman, D. W., and R. Mankau. 1979. Nematodes and microflora in the root rhizosphere of four desert shrubs.
Page 423 in J. L. Harley and R. S. Russell, editors. The
soil-root interface. Academic Press, London, England.

French, N., editor. 1979. Perspectives in grassland ecology.
Results and applications of the US/IBP grassland biome
study. Ecological Studies 32. Springer-Verlag, New York,
New York, USA.

Gerlach, S. H., and M. Schrage. 1969. Freilebende Nematoden
als Nahrung der Sandgarnele Crangon crangon. Oecologia
2:362-375.

Gould, W. D., R. V. Anderson, J. F. McClellan, D. C. Coleman,
and J. L. Gurnsey. 1979. Characterization of a paleosol:
its biological properties and effect on overlying soil
horizons. Soil Science 128:201-210.

Hanlon, R. D. G., and J. M. Anderson. 1979. The effects of
collembola grazing on microbial activity in decomposing
leaf litter. Oecologia 38:93-99.

Huhta, V., E. Ikonen, and P. Vilkamaa. 1979. Succession of
invertebrate populations in artificial soil made of sewage
sludge and crushed bark. Annales Zoologici Fennici 16:
223-270.

Huhta, V., and A. Koskenniemi. 1975. Numbers, biomass and
community respiration of soil invertebrates in spruce
forests at two latitudes in Finland. Annales Zoologici
Fennici 12:164-182.

Jones, F. G. W.   1957.   Soil population of beet eelworm
(Heterodera schactii Schm.) in relation to cropping. III.
Further experiments with microplots and with plots.
Nematologica 2:257-272.

Jones, F. G. W., and R. A. Kempton.   1978.   Population
dynamics, population models and integrated control.
Pages 333-361 in J. F. Southey, editor. Plant nematology.
Minstry of Agriculture, Fisheries and Food. Her Majesty's
Stationery Office, London, England.

Jones, F. G. W., D. W. Larbey, and D. M. Parrott.   1969.   The
influence of soil structure and moisture on nematodes,
especially Xiphinema, Trichodorus, Longidorus, and Heterodera
spp. Soil Biology and Biochemistry 1:153-165.

King, K. L., and K. J. Hutchinson.   1976.   The effects of sheep
stocking intensity on the abundance and distribution of
mesofauna in pastures. Journal of Applied Ecology
13:41-55.

Kira, T., Y. Ono, and T. Hosohawa.   1978.   Biological produc-
tion in a warm-temperate evergreen oak forest of Japan.
Japanese International Biological Program Synthesis,
Volume 18. University of Tokyo Press, Tokyo, Japan.

Kitazawa, Y.   1971.   Biological regionality of the soil fauna
and its function in forest ecosystem types. Pages 485-498
in Ecology and conservation No. 4. Productivity of forest
ecosystems. Proceedings of the Brussels Symposium 1969, UNESCO.

------.   1977.   Ecosystem analysis of the subalpine conifer-
ous forest of the Shigayama IBP area, central Japan.
Japanese International Biological Program Synthesis,
Volume 15. University of Tokyo Press, Tokyo, Japan.

Lane, C. L.   1975.   Forest stand conversion from hardwoods to
pines: effects on soil nutrients, microorganisms and

forest floor weight during the first seven years. Forest
Science 21:155-159.

Lee, J. J., and D. L. Inman. 1975. The ecological role of
consumers--an aggregated systems view. Ecology 56:1455-1458.

Macfayden, A. 1978. The role of the fauna in decomposition
processes in grasslands. Scientific Proceedings of the
Royal Dublin Society 6:197-206.

McKercher, R. B., T. S. Tollefson, and J. R. Willard. 1979.
Biomass and phosphorus contents of some soil invertebrates.
Soil Biology and Biochemistry 11:387-391.

Molloy, L. F., and L. C. Blakemore. 1974. Studies on a
climosequence of soils in Tussock grasslands. I. Introduction, sites, and soils. New Zealand Journal of Science
17:233-255.

Nicholas, W. L. 1975. The biology of free-living nematodes.
Clarendon Press, Oxford, England.

Paesler, F. 1946. Beitrag zur Kenntnis der im Dunger lebenden
Nematoden. Oesterreichische Zoologische Zeitschrift
1:87-128.

Palmisano, A. M., and T. Turchetti. 1975. Azione di un
nematode saprofago in rapporto all; antogonismo di Bacillus
subtilis (Cohn) Prazmowski verso Fusarium solani Mart.
Redia 56:117-134.

Paramonov, A. A. 1968. Plant-parasitic nematodes. Volume 1.
K. Skrjabin, editor. United States Department of Commerce,
Springfield, Virginia, USA.

Piearce, T. G., and M. J. Phillips. 1980. The fate of ciliates
in the earthworm gut: an in vitro study. Microbial Ecology
5:313-319.

Plowman, K. P. 1979. Litter and soil fauna of two Australian
subtropical forests. Australian Journal of Ecology 4:87-104.

Popovici, I. 1977. Distributia si dinamica populatiilor de

nematode din sol. Studii si Cercetari de Biologie, Seria
Biologie Animala 29:73-79.

Reichle, D. E. 1977. The role of soil invertebrates in
nutrient cycling. in U. Lohm and T. Persson, editors.
Soil organisms as components of ecosystems. Ecological
Bulletin (Stockholm) 25:145-156.

Sachs, H. 1950. Die Nematodenfauna der Rinderexkremente.
Zoologische Jahrbuecher 79:209-272.

Santos, P. F., J. Phillips, and W. G. Whitford. 1981. The
role of mites and nematodes in early stages of buried
litter decomposition in a desert. Ecology 62:654-663.

Sauerbeck, D. R., and B. G. Johnen. 1977. Root formation
and decomposition during plant growth. Pages 141-148 in
Soil organic matter studies. Volume 1. International
Atomic Energy Agency, Vienna, Austria.

Seinhorst, J. W. 1970. Dynamics of populations of plant
parasitic nematodes. Annual Review of Phytopathology
8:131-156.

Smolik, J. D. 1977. Effect of nematicide treatment on
growth of range grasses in field and glasshouse studies.
Pages 257-260 in J. K. Marshall, editor. The belowground
ecosystem: a synthesis of plant-associated processes.
Range Science Department Scientific Series No. 26.
Colorado State University, Fort Collins, Colorado, USA.

Sohlenius, B. 1980. Abundance, biomass and contribution
to energy flow by soil nematodes in terrestrial ecosystems.
Oikos 34:186-194.

Stanton, N. L., M. Allen, and M. K. Campion. 1981. The
effect of the pesticide carbofuran on soil organisms and
root and shoot production in a shortgrass prairie. Journal
of Applied Ecology 18:417-431.

Sykes, G. B. 1979. Yield losses in barley, wheat and potatoes

associated with field population of "large form" Longidorus
leptocephalus. Annals of Applied Biology 91:237-241.

Teal, J. M., and W. Wieser. 1966. The distribution and
ecology of nematodes in a Georgia salt marsh. Limnology
and Oceanography 11:217-222.

Twinn, D. C. 1974. Nematodes. Pages 421-465 in C. H.
Dickinson and G. J. F. Pugh, editors. The biology of
plant litter decomposition. Academic Press, London,
England.

Wasilewska, L. 1979. The structure and function of soil
nematode communities in natural ecosystems and agro-
cenoses. Polish Ecological Studies 5:97-145.

Wasilewska, L., H. Jakubczyk, and E. Paplinska. 1975. Production of Aphelenchus avenae Bastian (Nematoda) and
reduction of mycelium of saprophytic fungi by them.
Polish Ecological Studies 1:61-73.

Yeates, G. W. 1973. Abundance and distribution of soil
nematodes in samples from the New Hebrides. New Zealand
Journal of Science 16:727-736.

------. 1974. Studies on a climosequence of soils in tussock
grasslands. 2. Nematodes. New Zealand Journal of Zoology
1:171-177.

------. 1979. Soil nematodes in terrestrial ecosystems.
Journal of Nematology 11:213-229.

------. 1980a. Populations of nematode genera in soils under
pasture. III. Vertical distribution at eleven sites.
New Zealand Journal of Agricultural Research 23:117-128.

------. 1980b. Relation of generic nematode populations to
soil and plant parameters. Journal of Nematology 12:242-
243.

------. 1981. Soil nematode populations depressed in the
presence of earthworms. Pedobiologia 22:191-195.

PARAMETERS OF THE NEMATODE CONTRIBUTION TO ECOSYSTEMS

Diana W. Freckman

It is difficult to quantify the role of invertebrates in a soil ecosystem, and nematodes are no exception. Nematodes are a diverse group of roundworms dependent on a thin film of water around soil particles for their movement and completion of their life cycle. They are abundant and ubiquitous, occurring in all soils, ranging from arctic to desert origins. However, because of their microscopic size, clumped distribution and trophic variation, it is often difficult to determine their contribution to the dynamics of the soil ecosystem. The purpose of this paper is to discuss the procedures used to quantify energy and material flow through nematodes. For additional information consult Freckman and Baldwin (in press), Nicholas (1975), Norton (1978), Sohlenius (1980), and Yeates (1979).

SAMPLING

It is impossible to count each nematode in the soil, so we are dependent on some sampling procedures to estimate nematode population density and community structure. An adequate and representative sampling program is one of the most critical factors in obtaining quantitative nematode data (Bird 1978, Goodell and Ferris 1980, Goodell's chapter, this volume, Proctor and Marks 1975, Southey 1970, Wasilewska and Paplinska 1976). The object of sampling and extraction is to collect and extract a sample that accurately reflects the nematode population and community structure in the soil.

Nematode distribution varies both horizontally and vertically and it is influenced by many factors including plant

roots, litter and soil type. Migration into an area is not
of major importance, although one phytoparasitic species,
Meloidogyne incognita, can migrate 120 cm vertically to host
roots (Johnson and McKeen 1973, Prot and Van Gundy 1980). Of
more importance is the association of nematodes to their food
sources, plant roots, soil microfauna and microflora. By
sampling only scattered plants which have high accumulations
of litter an overestimate can be made of numbers and total
nematode biomass for a given ecosystem. Similarly, shallow
sampling of a plant with a deep root system can underestimate
plant parasitic nematode numbers and biomass.

Nematode distribution varies with soil type. For example,
a nematode species may be more abundant in a light textured
sandy soil than in a heavy textured clay soil. Goodell and
Ferris (1980) found that nematode species and abundance was
significantly affected by soil type in a 7 ha alfalfa field.
If the plot size had represented only the clay portion of the
field, estimates of nematode abundance and community structure would have been distorted.

Goodell and Ferris (1980) showed that it is more econom-
ical and more efficient to take several cores and combine
them into one sample rather than using one core per one sample. The Society of Nematologists (Barker et al. 1978)
listed recommendations for: (1) the diameter, number and
depth of cores needed to provide an adequate sample; (2)
representative patterns of sampling to provide reliable data
on population densities; (3) the sampling frequency to reflect
population densities at critical stages of life cycle and,
(4) proper handling and storage of samples. Concerning
storage of samples, it is important to remember that exposure
of samples to temperatures greater than 30 C, even for a

short time can kill certain species and thus interfere with recovery and identification.

The next consideration in sampling is required frequency of sampling, which depends on the purpose of the study. For nematode community analyses involving spatial and seasonal distribution, samples should be collected at least monthly for a year. This is because of seasonal variation in the environmental factors such as soil temperature and moisture which influence nematode biology. In addition, some nematodes have only 1 generation per year, whereas others have up to 8 generations per year, thus the structure of the community may vary with time.

EXTRACTION

Extraction methods vary in efficiency and ability to extract different species equally. Nematode extraction methods either (1) extract nematodes in water (semi automatic elutriator, Seinhorst elutriator, Oostenbrink elutriator, Cobb's sieving and decanting, Baermann funnel, mist chamber), or (2) return them to water after flotation (sugar flotation sieving, centrifugal flotation). These methods are used because nematodes are essentially aquatic organisms. The Baermann funnel or its adaptations extracts only actively moving adults or larvae from roots or soil. The ideal method would extract representative weights, sizes and life stages of each nematode trophic group. Several extraction methods should be tested and evaluated prior to a study. In a comparison of extraction methods for desert soils (Freckman et al. 1978), samples taken during the hot summer months and placed on a Baermann funnel for 48 h yielded high numbers of juveniles when compared to Cobb's sieving and decanting and sugar flotation sieving techniques, because eggs had hatched

and the nematode counts reflected potential population or biomass, and not the population in the soil at the time of
sampling.

The Baermann funnel is a commonly used technique which is
discussed here to illustrate the necessity of recognizing a
technique's limitations. It usually has a low extraction
efficiency due to: (1) variation in nematode motility with
species (microbivores are more active than ectoparasitic
phytophages) and with temperature; (2) asphyxiation: nematodes remaining too long at the bottom of the funnel may die
due to lack of oxygen and increased microbial activity; (3)
excessive water over soil or too much soil makes a poor recovery and, (4) use of deionized water reduces nematode extraction. See Barker et al. (1978), Ayoub (1980) for limitations of other methods. Many ecology studies have largely
ignored the extraction of endoparasites of roots, because
this often requires a separate method (mist chamber) with a
longer incubation period. Regardless of the ecosystem
being studied, it is important to know if both endoparasites
and ectoparasites will comprise the phytophagous trophic
group and to use the appropriate extraction techniques.

It is essential that the extraction efficiency is determined for each soil ecosystem study and reported in the
literature for comparative purposes and because different
methods favor certain species. In addition, densities should
be cited with their standard deviation. There is considerable
discussion among nematologists regarding method of efficiency
determination. Seeding of the soil, litter, or lichens with
a known quantity of nematodes is a common method. When adding
the nematodes to the soil prior to extraction, it is important to use nematodes of different sizes and representing
different trophic groups. Cultures of different trophic

species can be obtained from many agricultural and ecology
centers. (For information of sources of nematode cultures,
contact Chairman, Education Committee, Society of
Nematologists).

IDENTIFICATION AND TROPHIC GROUPS
Many nematodes can be identified to genus level using a dis-
secting scope (45x-100x) and transmitted light. Species
identification usually requires carefully prepared slides of
adults and examination under oil immersion. The specimens
must be inactivated (with fixative or heat) for the time necessary to observe them. A mass collection (water containing
large numbers of nematodes) can be preserved by adding an
equal volume of boiling 10% formalin to the water. (See
Ayoub 1980, Freckman and Baldwin, in press, Goodey 1963, and
Southey 1970).

Appropriate classification groups must be determined before
beginning identification. It is not enough to list nematodes
with stylets as plant feeding and those without stylets as
omnivores or microbial feeders. Perferably, the fauna can
often be listed as genera within trophic groups. For
example, the biology of a nematode identified as a "Mononchus"
is most assuredly a predator. Useful keys for nematodes include Goodey (1963), Freckman and Baldwin (in press), Mai
and Lyon (1975) and Thorne (1961).

Nematodes feed on living protoplasm and so none are known
to be saprophagic. The trophic groups commonly used (based on
the schemes of Banage (1963), Wasilewska (1971), and Yeates
(1979) are:

1. Phytophages (can be migratory or sedentary and are
    obligate parasites)

    a) endoparasites (feed within roots)

    b) ectoparasites (feed on surface of roots)

2. Microbivores (feed on bacteria and other microflora)

3. Fungivores (feed on fungal mycelium)

4. Omnivores (consume fungi, bacteria, algae, protozoans, and rotifers)

5. Predators (feed on other nematodes, enchytraeids, tardigrades, and protozoa)

A specific trophic group is often inferred from the morphology of the nematode stoma. Nematodes with protrusible hollow stylets are usually plant feeders, but could also be fungivores, predators, or parasites of invertebrates. Fungivores usually have stylets, microbivores an open, unarmed stoma, and predators have small axial teeth and/or an odontostyle. These trophic groups are not definitive (Nicholas 1975). Our lack of knowledge concerning nematodes and their trophic relationships was illustrated by Sohlenius (1968) who examined competition for nutrients between two bacterial feeders. He discovered that one of the bacterial feeders consumed the other. The literature should be checked for species food association before starting a trophic based study and feeding studies should be more frequently undertaken.

NEMATODE WEIGHT

Average individual weight is probably the most important parameter in an ecology study. It influences biomass, respiration, and production. It is difficult to weigh an individual nematode because of the small size (0.3–2.5 mm) and because of any soil or debris associated with the nematode cuticle after extraction. Also, because 75–80% of their body weight is water, any additional water would have to be accounted for. Due to these difficulties, other techniques were developed.

Before 1957, there were two methods used to calculate the weight of nematodes. Nielsen (1949) considered a nematode as cylindrical, with the radius determined at the esophagus. The use of this rapid formula for a cylinder resulted in only an approximation of nematode weight. Volz (1951) made plastic scale models of nematodes. These were placed in water and the specific gravity and volume determined. This was an inefficient and laborious method, requiring considerable skill and not suitable for large numbers of weight determinations.

In 1956, Andrassy made drawings of nematodes using a camera lucida. The length and width was measured and the nematode divided into either three truncated cones with a cone for the tail, or into a cylinder, truncated cone and a cone for the tail. This mathematical method required about 24-27 different calculations and it is based on several assumptions. For example, tail shapes were considered to be of 2 types, conical or rounded. Other tail types were visually rounded off if the weight beyond a certain point was considered negligible. The weights of cuticular appendages were not determined. Because this was a lengthy method, Andrassy developed the following formula:

$$W = \frac{(w^2)\ (\ell)}{(1.6)\ (10^6)} \tag{1}$$

where:

W = weight in μg

w = width of the nematode at the widest (not the the vulva) point.

ℓ = total length of the nematode in μm.

Tails not resembling a cone were extrapolated to that shape. Andrassy's comparisons of measurements from the mathematical

27 step formula and the shortened version showed that variation was only $\pm$ 5% using the latter.

Since 1956, Andrassy's shortened formula has been used to determine individual nematode weight. Usually a conversion factor of 0.24 dry weight (^{dw}) is used (Yeates 1979). Croll and de Soyza (1980) determined caloric values for a wide range of nematodes and found a mean of 5.095 kcal/g dw, although 2.152 cal/mg fresh weight (fw) is used according to Yeates (1979).

L. Duncan and Freckman (unpublished) retested the mathematical formula against the shortened formula with different sizes and tail shapes of nematodes. The mathematical formula used was more detailed than that of Andrassy. Length measurements were the actual tail length, not extrapolated in case of roundness. A formula other than for a cone was used if the tail was rounded. At least 25 width measurements were taken. The nematode was then divided into numerous truncated cones, cylinders and cones as determined by width measurements. Variations between these laborious mathematical calculations and Andrassy's short formula was found to be 5-31%. This is alarming, considering that the equations for determining biomass, respiration, and production are all dependent on the short Andrassy formula. This area needs further examination.

NEMATODE RESPIRATION

The same reasons that limit nematode weight measurements, also make difficult determination of individual nematode oxygen consumption. Respiration for soil nematodes is calculated by multiplying nematode fresh weight by estimated respiration rate over time and adjusting to field temperature. A commonly used formula is that of Klekowski et al. (1972):

$$R = 1.4G^{0.72} \qquad\qquad (2)$$

where:

$R = \mu l\ O_2$ consumed $\cdot 10^{-3}$/individual/h

G = fresh weight in μg

In this case, respiration was measured for individual nematodes at 20 C using a Cartesian diver. The nematodes represented a variety of different trophic groups, sizes, and life
stages. Soil temperature/respiration relationships can be
found using Krogh's curve (Winberg 1971) with Q value cor-
rections according to Duncan and Klekowski (1975). Wasilewksa
(1971) used different respiration rates for different feeding
groups. A calorific equivalent of 4.8 cal/ml O_2 is used
(Yeates 1973). Community metabolism is found by multiplying
(number of nematodes) (R) (number of hours per month).
Nielsen (1961) had used ml O_2/g fresh weight but this is now
recognized as an underestimation of nematode respiration.
Klekowski and Paplinska (1974) found that juvenile respiration was 2 times higher than adult respiration and presented
a formula to account for respiration of the juvenile stages.

    Another consideration when measuring nematode respiration
is cryptobiosis. Under severe environmental stress, soil
nematodes can enter into a reversible inactive state in which
there is no metabolism (Demeure and Freckman 1981). Some of
the environmental conditions and corresponding cryptobiotic
states are: freezing-cryobiosis, desiccation-anhydrobiosis,
and lack of oxygen-anoxybiosis. The duration of this cryptobiotic period should be taken into consideration when working in a soil ecosystem that is subject to environmental extremes. For example, in a weekly sampling study at Rock

Valley, Nevada (Freckman 1978, Freckman and Mankau, in press)
when soil moisture went below 2.5% the nematodes entered into
anhydrobiosis and had no measurable oxygen consumption (Fig. 1).

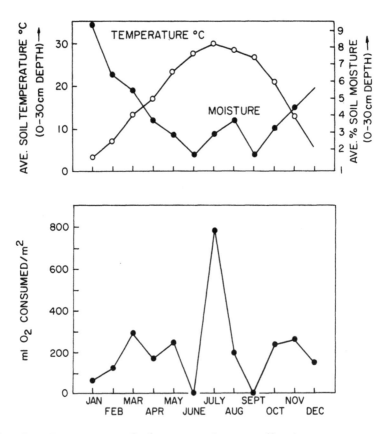

Fig. 1.   Parameters of the nematode contribution to ecosystems.

Previously, soil ecologists (Nielsen 1961, Jones 1975) sub-
tracted nematode respiration for the month of the drought
period. From soil temperature and moisture data for each
sampling depth and at each sampling time a better estimate of
the duration of the anhydrobiotic period can be determined.
This knowledge, plus the use of the anhydrobiotic technique
(Freckman et al. 1977) enables a more precise estimation of
the nematode inactive period.

Other sources of error in estimations of nematode respiration which are related to weight measurements include mea-
suring only a few (<25) nematodes per trophic group and mea-
suring only adult stages. Sohlenius (1979, 1980) suggested
that using frequency of hours with different soil temperatures
within a month would obtain a better temperature/respiration
correction than using mean monthly temperature values.

PRODUCTION
Direct and independent field measurements of production (P)
and respiration (R) are impossible to take, therefore, these
parameters are usually calculated by multiplying individual
data on production and respiration, respectively, by population
numbers or biomass. Englemann's (1966) equation was used to
calculate production until 1970. His equation was derived
from 8 measurements. McNeill and Lawton's (1970) equation for
short-lived poikliotherms had about 50 measurements including
marine and aquatic fauna not considered by Englemann and is
more widely used, although still controversial (Sohlenius
1980). This formula is also dependent on nematode weight:

$$\log P = 0.8262 \log R - 0.0948 \qquad (3)$$

where:

$$R = \text{amount of kilocalories used for respiration}/m^2/yr$$

Consumption may then be estimated using the assumption of Kitazawa that nematode consumption is twice as high as assimilation, i.e., both maintenance and production costs (Wasilewska 1974).

DISCUSSION

Body width and length, and respiration were examined by sensitivity analyses (Table 1) to see what effect underestimation of the parameter by 5% would have on production.

TABLE 1. Sensitivity analysis of some of the parameters used to estimate production. Nematode width, length and respiration have been underestimated by 5%.

Nematode width	Weight length	Respiration	Production ($kcal/m^2/yr$) Extraction Efficiency			% change*
			100%	75%	33%	
−	−	−	2.151	1.696	.860	
+	−	−		1.598	.810	6
−	+	−		1.647	.836	3
−	−	+		1.626	.825	4
+	−	+		1.531	.777	10
−	+	+		1.579	.801	7

+ indicates a 5% underestimation
− indicates a perfect measurement
* independent of extraction efficiency

These errors could occur if mistakes were made in width measurements, or if only adults were measured. Extraction was compared at a 75% and a 33% efficiency. Nematode width was

20 μm and length was 500 μm, giving a nematode weight of
.125 μg. It was assumed that there were 1×10^6 nematodes/
m^2/0-10 cm soil depth. Results indicate extraction efficiency
affects production more than any of the other variables analyzed. If compared to an unrealistic 100% efficiency the
percent change would be 21 and 60% for a 75 and 33% extraction
efficiency. It is interesting to note that the width measurement, which is a squared factor in Andrassy's formula,
also has a major effect on production estimates, whether alone
or in combination with changes in respiration. The analyses
indicate that the error that might be encountered when using
these formulae to estimate production is not great unless
combined with errors such as sampling and identification.

Application of the correct ways to increase the accuracy
of estimating nematode structure and function are required to
make this area of soil ecology a more exacting science.

LITERATURE CITED

Andrassy, I. 1956. Die rauminhalts und gewichtsbestimmung
    der fadenwurmer (Nematoden). Acta. Zoologica Academi
    Sciences Hungary 2:1-15.

Ayoub, S. M. 1980. Plant nematology. An agricultural
    training aid. Department of Food and Agriculture,
    Sacramento, California, U.S.A.

Banage, W. B. 1963. The ecological importance of free-living
    nematodes with special reference to those of moorland soil.
    Journal of Animal Ecology 32:133-140.

Barker, K. R. (Chairman) 1978. Determining nematode popu-
    lation responses to control agents. Pages 114-125 in
    E. I. Zehr, editor. Methods for evaluating plant fungi-
    cides, nematicides and bactericides. American Phytopatho-
    logical Society, St. Paul, Minnesota, U.S.A.

Bird, G. W. 1978. Nematode detection. Michigan State
University Agricultural Facts, No. 9; Cooperative Extension Service, East Lansing, Michigan, U.S.A.

Croll, N. A., and K. de Soyza. 1980. Comparative calorie
values of nematodes. Journal of Nematology 12:132-135.

Demeure, Y., and D. W. Freckman. 1981. Recent advances in
anhydrobiosis. Pages 205-226 in B. M. Zuckerman and R. A.
Rohde, editors. Plant parasitic nematodes, Volume III.
Academic Press, New York, New York, U.S.A.

Duncan, A., and R. Z. Klekowski. 1975. Parameters of an
energy budget. Pages 97-147 in W. Grodzinski, R. Z.
Klekowski, and A. Duncan, editors. Methods for ecological
bioenergetics. IBP Handbook 24. Blackwell Scientific
Publications, Oxford, England.

Englemann, M. D. 1966. Energetics, terrestrial field studies
and animal productivity. Advances in Ecological Research
3:73-115.

Freckman, D. W. 1978. Ecology of anhydrobiotic soil nema-
todes. Pages 345-357 in J. Crowe and J. Clegg, editors.
Dried biological systems. Academic Press, New York, New
York, U.S.A.

Freckman, D. W., and J. G. Baldwin. Soil nematoda. in D. L.
Dindal, editor. Soil biology. Wiley-Interscience, New
York, New York, U.S.A. (in press).

Freckman, D. W., and R. Mankau. Distribution, abundance
and productivity of soil nematodes in a northern Mojave
desert site. Journal of Arid Environments (in press).

Freckman, D. W., R. Mankau, and H. Ferris. 1975. Nematode
community structure in desert soils: nematode recovery.
Journal of Nematology 7:343-346.

Freckman, D. W., D. T. Kaplan, and S. D. Van Gundy. 1977.
A comparison of techniques for extraction and study of

anhydrobiotic nematodes from dry soils. Journal of
Nematology 9:176-181.

Goodell, P. B., and H. Ferris. 1980. Plant parasitic nematode
distributions in an alfalfa field. Journal of Nematology
12:136-140.

Goodey, J. B. 1963. Laboratory methods for work with plant
and soil nematodes. Technical Bulletin No. 2. Ministry of
Agriculture, Fisheries and Food. Her Majesty's Stationery
Office, London, England.

Johnson, P. W., and C. D. McKeen. 1973. Vertical movement
and distribution of Meloidogyne incognita (Nematoda) under
tomato in a sandy loam greenhouse soil. Canadian Journal
of Plant Science 53:837-841.

Jones, F. G. W. 1975. The soil as an environment for plant
parasitic nematodes. Annals of Applied Biology 79:113-139.

Klekowski, R. Z. L. Wasilewska, and E. Paplinska. 1972.
Oxygen consumption by soil inhabiting nematodes.
Nematologica 18:391-403.

Mai, W. F., and H. H. Lyon. 1975. Pictorial key to genera
of plant-parasitic nematodes. Cornell University Press,
Ithaca, New York, U.S.A.

McNeill, S., and J. H. Lawton. 1970. Annual production and
respiration in animal populations. Nature 225:472-474.

Nicholas, W. L. 1975. The biology of free-living nematodes.
Clarendon Press, Oxford, England.

Nielsen, C. O. 1949. Studies on the soil microfauna. II.
The soil inhabiting nematodes. Natura Jutlandica 2:1-126.

-------, 1961. Respiratory metabolism of some populations
of enchytraeid worms and freeliving nematodes. Oikos
12:17-35.

Norton, D. C. 1978. Ecology of plant parasitic nematodes.
John Wiley and Sons, New York, New York, U.S.A.

Proctor, J. R., and C. F. Marks. 1975. The determination of normalizing transformations for nematode count data from soil samples and of efficient sampling schemes. Nematologica 20:395-406.

Prot, J. C., and S. D. Van Gundy. 1981. Effect of soil texture and the clay component on migration of *Meloidogyne incognita* second-stage juveniles. Journal of Nematology 13:217-219.

Sohlenius, B. 1968. Studies of the interactions between *Mesodiplogaster* sp. and other rhabditid nematodes and a protozoan. Pedobiologia 8:340-344.

------------ 1979. A carbon budget for nematodes, rotifers, and tardigrades in a Swedish coniferous forest soil. Holarctic Ecology 2:30-40.

------------ 1980. Abundance, biomass, and contribution to energy flow by soil nematodes in terrestrial ecosystems. Oikos 34:186-194.

Southey, J. F. 1970. Principles of sampling for nematodes. Pages 1-4 in J. F. Southey, editor. Laboratory methods for work with plant and soil nematodes. Ministry of Agriculture, Fisheries and Food. Her Majesty's Stationery Office, London, England.

Thorne, G. 1971. Principles of nematology. McGraw-Hill Book Company, New York, New York, U.S.A.

Volz, P. 1951. Untersuchungen uber die mikrofauna des waldbodens. Zoologische Abt. Jahrbuecher Syst. Geogr. Tiere, 79:514-566.

Wasilewska, L. 1971. Nematodes of the dunes in the Kampinos forest. II. Community structure based on number of individuals, state of biomass, and respiratory metabolism. Ekologia Polska 19:651-688.

Wasilewska, L., and E. Paplinska. 1976. Method of soil
    sampling and estimation of numbers of biomass and the
    community structure of soil nematodes. Ekologia Polska
    24:593-606.

Winberg, G. C. 1971. Methods for the estimation of production
    of aquatic animals. Academic Press, London, England.

Yeates, G. W. 1973. Nematoda of a Danish beech forest. II.
    Production estimates. Oikos 24:179-185.

------------ 1979. Soil nematodes in terrestrial ecosystems.
    Journal of Nematology   11:213-228.

THE ROLE OF NEMATODES IN DECOMPOSITION
IN DESERT ECOSYSTEMS

W. G. Whitford, D. W. Freckman, P. F. Santos,
N. Z. Elkins, and L. W. Parker

Although nematodes are numerous in the soils of most ecosystems, their contribution to soil respiration is relatively
small (Sohlenius 1980). Thus, judged on the basis of energy
flow through an ecosystem, nematodes would have to be considered a minor component of such systems. However, the contribution of a group of organisms to the functioning of a
system cannot be judged solely on the basis of the quantity
of energy processed by that group. Chew (1974) suggested
that while consumers may be relatively unimportant in energy
flow, they may play important roles as rate regulators. In
this paper we review our findings to date on the role of
nematodes in desert soil ecosystems and show how various
guilds of free-living nematodes apparently regulate rates of
decomposition, thus eventually affecting nutrient cycling
processes.

Free-living nematodes, especially bacteriophagic nematodes,
make up the bulk of the desert soil fauna (Freckman and Mankau
1977). Free-living nematodes are indirectly dependent upon
dead plant material for their energy and nutrients via bacteria
and fungi. Therefore, the distribution and relative abundances
of nematodes in deserts is a function of the distribution of
plant litter as documented by Freckman and Mankau (1977).

There are two fates for litter in deserts: (1) accumula-
tion under shrubs or in shrub clumps where deposited by wind
and water, or (2) buried by wind transported soil in depres-
sions made by animals and buried along small water courses
behind obstructions. Another source of organic matter which

can serve as a focus for free-living nematodes are the roots of ephemeral plants. In deserts ephemeral plants are an important component of the flora. When such plants complete seed set, they die leaving the dead roots to decompose in situ. This material then becomes available for decomposition by the soil flora and fauna. Ephemerals provide a pulse input of organic matter to the desert soil ecosystem. It is not surprising that Freckman and Mankau (1977) reported low percentages of root parasites. Growth of ephemeral plants occurs in irregular and brief pulses and growth of most perennial plants is tied to soil moisture and hence is sporadic. The sporadic activity of roots in desert soils undoubtedly reduces the diversity of root parasites and affects the populations of those species which do successfully parasitize roots of desert plants. However, these relationships must remain speculative until these hypotheses are tested.

With such an array of sources of energy and nutrients in a variety of micro-environments, it is clear that understanding the role of nematodes in such a system is a complex task.

There are two basic approaches which have been used to study soil processes: (1) microcosms, and (2) field studies. Microcosms provide controlled microclimates in which a variety of combinations of soil flora and fauna can be examined. However, at some stage, studies must be conducted in natural environments preferably before the investigator's ideas about what might be happening in nature becomes too firmly fixed (Brock 1971). Field studies have the advantage of incorporating natural microclimate variation involving the complete soil flora and fauna, but require more time and do not allow examination of interactions between individual taxa.

We chose to do field studies in order to obtain assessments of the relative importance of various components of the soil

biota in decomposition processes. We used selective chemical
inhibitors in our studies. Regardless of the technique chosen
for a given experiment, it is important to be cognizant of
the limitations of that technique. Use of pesticides, fungicides and/or nematicides poses problems of inhibition of nontarget organisms, gradual loss of toxicity to target organisms
and/or failure to completely eliminate or inhibit all taxa of
a target group. The advantage of using such inhibitors is
that experiments can be done under field conditions providing
a measure of the responses to natural environmental fluctuations and to the entire spectrum of biological interactions.
While lacking the precision of the more elegant laboratory
studies, the system responses are readily applicable to the
unperturbed ecosystems, especially if the limitations discussed above are well known. Of the pesticides available for
use at the time, chlordane, according to the literature, had
the lowest nontarget (nonarthropod) effects. The fungicides
we used undoubtedly inhibited some if not all of the soil
bacteria and did not suppress all groups of fungi. The
nematicide, Nemagon®, was a broad-spectrum soil sterilant and
responses of the system to the nematicide treatment had to be
so interpreted.

In all of the studies of buried litter and roots, we used
similar techniques. Selected quantities of litter or roots
(20-30 g) were placed in fiberglass mesh bags. The bags were
then soaked in water plus wetting agent, or solutions of insecticide or fungicide depending upon the assigned treatment.
The bags were buried in the soil at approximately 10 cm depth
and retrieved at designated intervals. Microarthropods were
extracted from the material remaining in the bags by Tullgren
funnels into water and counted and identified immediately
after extraction (see Santos and Whitford 1981 for details).

Material in bags selected for nematode extractions was placed in a blender with water and blended for 10 sec. That mixture was washed through a series of filters and the material caught on 44-μm (325-mesh) and 38-μm (400 mesh) screens washed onto a cotton filter. This is a modification of the Cobb sieving and Oostenbrink cotton wool filter techniques (Nicholas 1975). Nematodes were assigned to trophic groups based on esophageal structures (Yeates 1973). Bacteria were counted directly using fluorescent stain (Babiuck and Paul 1970) and hyphal lengths measured using a modification of the technique of Olson (1950).

Our initial studies (Santos and Whitford 1981) were designed to examine the relationships between various groups of soil biota during the initial stages of decomposition. Santos and Whitford (1981) had shown that the decomposition of buried creosotebush (<u>Larrea</u> <u>tridentata</u>) leaf litter was most rapid in the first month in the soil and that inhibition of microarthropods and/or fungi reduced the rate of decomposition (Fig. 1). This study demonstrated that decomposition

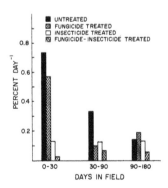

Fig. 1. The effects of selected biocides on decomposition rates of creosotebush litter buried in fiberglass mesh bags in a Chihuahuan desert ecosystem, May through October.

rates were higher in the first 30 d than during the remainder of the decomposition process. Santos and Whitford (1981) also found large numbers of nematodes in litter bags from which microarthropods had been excluded. These relationships raised questions about the role of nematodes in the decomposition process.

In a study designed to examine decomposition and changes in soil biota in the initial 30 d after burial of creosote-bush leaf litter, we (Santos et al. 1981) found that exclusion of the tydeid mites by chlordane resulted in increased density of cephalobid nematodes and decreased density of bacteria (Table 1). In cultures of tydeid mites and nematodes, the tydeid mites fed voraciously on the eggs and all life stages of the nematodes. Based on these data we suggested that elimination of the tydeid mites released the nematode population from predation pressure. The larger numbers of nematodes overgrazed the bacteria, thus reducing the rate of decomposition. This was a short lived relationship because by day 25 the densities of bacteriophagous nematodes decreased and fungi and fungiphagous nematodes began to increase.

These data suggest that predatory mites maintain nematode grazer populations at levels which presumably increase microfloral activity thus enhancing decomposition. Hanlon and Anderson (1979) demonstrated that low densities of fungal grazing Collembola stimulated microbial respiration but high densities reduced respiration. Their studies and ours suggest that unless controlled by predators, grazers are detrimental to decomposition.

If the balance between predatory mites and free-living nematodes affected grazing on primary decomposers (bacteria and fungi) in the early stages of leaf litter in a Chihuahuan desert ecosystem, we asked, how generalized is that relation-

TABLE 1. Percent Organic Matter Loss ± One Standard Deviation from Buried Litter Bags (N=5) and Numbers ± One Standard Deviation of Bacteria (N=3), Tydeid Mites (N=5), and Bacteriophagous Nematodes (N=3) Per g of Litter in Buried Insecticide Treated (IT) and Untreated (NT) Creosotebush (Larrea tridentata) Litter Bags. Data are Reported for the Time Periods Following Burial.

	Day 10		Day 20	
	NT	IT	NT	IT
Organic matter weight loss	20.7 ± 2.4%	5.6 ± 2.5%	23.3 ± 1.3%	8.2 ± 2.4%
Bacteria	3.3×10^6	2.0×10^6	1.0×10^7	1.6×10^6
Bacteriophagous nematodes	4.6 ± 2.2	25.0 ± 9.4	1.2 ± .2	9.3 ± 3.4
Tydeid mites	3.4 ± 1.0	0	4.7 ± .9	0

ship and does it pertain to the later stages of decomposition?
To answer these questions we conducted experiments using the
same inhibitors in the North American hot deserts and a semi-
arid transitional ecosystem in S.E. New Mexico. We also
examined the soil biota and decomposition of buried leaves,
stems and roots of a spring ephemeral (Lepidium lasiocarpum)
over the summer growing season.

Weight losses and mite and nematode numbers from litter
bags containing 10 g shinnery oak (Quercus harvardii Rydb.)
from the litter decomposition study in the transitional
shinnery oak ecosystem near Carlsbad, New Mexico are summarized in Table 2. As in the creosotebush buried litter, the
insecticide treated bags had significantly higher densities of
free-living nematodes than did untreated controls. In this
system mesostigmatid mites (Rhodacaridae and Laelapidae) were
the dominant predators. Using mites and nematodes (both
bacteriophagous and fungiphagous nematodes) extracted from

TABLE 2. Percent Organic Matter Loss ± One Standard Deviation
(N=7) from Buried 10 g Shinnery Oak (Quercus harvardii Rydb.)
Litter Bags and Numbers ± One Standard Deviation of Free-Living
Nematodes (N=3) and Predaceous Mites (N=7) Extracted from
Buried Insecticide Treated (IT) and Untreated (NT) Bags Buried
for 84 d at a Chihuahuan Desert Site Near Carlsbad, New Mexico

	NT	IT
Organic matter weight loss	18.10 ± .80%	7.64 ± .97%
Free-living nematodes	1140 ± 210	7811 ± 465
Predatory mites (Gamasina)		
Ascidae	4 ± 7	0
Laelapidae	39 ± 14	0
Rhodacaridae	315 ± 69	0

litter bags at this site, we conducted a simple feeding experiment. Twelve agar plates were prepared containing BBL®
beef extract and peptone nutrient broth (10 g $\cdot$ litter^{-1}) and
were air dried for 2 d. Fifty randomly selected nematodes,
suspended in 1 ml distilled water, were placed on each of the
12 plates. Ten Rhodacarid mites were added to 6 of the plates.
All plates were covered, placed in the dark for 48 h and then
examined and counted. Less than 10% of the original number of
nematodes remained alive on the plates containing predaceous
mites while more than 95% remained alive on plates without
mites.

The relationship between free-living nematodes, microarthropods and decomposition rates holds not only for the arid
regions of southern New Mexico but also for other North
American hot deserts (Table 3). Although there were species
differences in the predatory microarthropods, i.e., tydeids
and gamasids, inhibition of the predatory mites resulted in
increases of several orders of magnitude of nematodes which
graze on bacteria or fungi. The generality of this relationship strengthens the argument that while moderate grazing on
bacteria and fungi by nematodes enhances decomposition, high
densities of grazers produce marked reductions in decomposition.

Free-living nematodes appear to play very different roles
in the decomposition of dead roots when compared to the decomposition of above ground parts of the same annual plant
species. In our studies of pepperweed L. lasiocarpum we
found that during the longer exposure in the soil, fungi and
fungal based food chains were more important in decomposition
of litter than the bacteria and bacterial based food chains
(Table 4). This suggests that bacteria grow rapidly on low
molecular weight, readily degradable carbohydrates of leaf

TABLE 3. Percent Organic Matter Loss ± One Standard Deviation (N=5) from Buried 20 g Creosotebush Litter Bags and Numbers ± One Standard Deviation of Free-Living Nematodes (N=3) and Mites (N=5) Per Bag in Insecticide Treated (IT) and Untreated (NT) Bags Buried for 60 d in Different North American Deserts.

	Sonoran Desert Casa Grande, AZ		Coloradan Desert Glamis, CA		Mojave Desert Boulder City, NV	
	NT	IT	NT	IT	NT	IT
Organic matter weight loss	25.6 ± 2.4%	14.4 ± 3.9%	26.5 ± 4.8%	15.3 ± 3.1%	43.4 ± 5.7%	20.7 ± 5.4%
Free-living nematodes	164 ± 61	5230 ± 5117	266 ± 113	1338 ± 256	945 ± 245	2760 ± 758
Predatory mites						
Prostigmata						
Tydeidae	144.6 ± 175.3	0	9.6 ± 45	0	21.4 ± 6.8	0
Raphignathidae	20 ± 10.6	0	0	0	0	0
Gamasina (Mesostigmata)						
Arctacaridae	0	0	7.4 ± 4.8	0	37 ± 31	0
Other Gamasina	0	0	14.6 ± 10.5	0	18.4 ± 13	0

TABLE 4. Results of the Stepwise Regression Analysis of Populations of Soil Biota Related to Percent Weight Remaining in Litter Bags Containing Pepperweed, Lepidium lasiocarpum Roots and Above Ground Parts (Litter). DFMA-Detrivore-Fungivore Microarthropods, FMA-Fungiphagous Microarthropods, FNEM-Fungiphagous Nematodes, BNEM-Bacteriophagous Nematodes, NMA-Nematode Predatory Microarthropods. Minus (-) Sign Indicates Enhancement of Decomposition. Plus (+) Sign Indicates Reduction of Decomposition.

Treatment	Roots Variable Entered	r^2	Litter Variable Entered	r^2
Untreated	= -DFMA	.54	= -FUNGI - FNEM	.67
	-DFMA - FMA	.65	-FUNGI - FNEM - FMA	.82
			-FUNGI - FNEM - FMA - BACTERIA	.89
			-FUNGI - FNEM - FMA - BACTERIA	.93
			- FNEM - FMA - BACTERIA	.90
			- FNEM - FMA - BACTERIA + BNEM	.94
			- FNEM - FMA - BACTERIA + BNEM - NMA	.97
Insecticide treated	= -FUNGI	.70	= -FUNGI - FNEM	.73
	-FUNGI - FNEM	.98	-FUNGI - FNEM	.84
	-FUNGI - FNEM - BNEM	.99	-FUNGI - FNEM + BNEM	.91
	-FUNGI - FNEM - BNEM -BACTERIA	.99		
Fungicide insecticide treated	= -BNEM	.72	= -BNEM - FNEM	.89
	-BNEM - FNEM	.78	-BNEM - FNEM	.92

litter which accounts for the initial rapid weight loss. The
fungi begin to grow using the more recalcitrant substrates
and the process shifts to a fungal based food chain.

In untreated roots, decomposition was correlated with microarthropods we classified as detritivores or fungivores, and
nematode densities did not alter the stepwise regression
(Table 4). In the inhibitor treated roots, decomposition
was correlated with fungi, bacteria and free-living nematodes
(Table 4).

In contrast the changes in fungiphagous nematode densities
accounted for between 11% and 14% of the variation in decomposition of the pepperweed above ground parts (Table 4). In the
Lepidium litter as in the experiments described previously,
there was a reciprocal response of bacteria and bacteriophagous
nematodes in that bacterial densities increased when densities
of bacteriophagous nematodes decreased (Table 4).

Fungal grazers apparently play an important role as regulators of decomposition. Over long time periods, the grazing
by fungiphagous nematodes appears to be little affected by the
absence of microarthropod predators. These data suggest that
factors other than predation are the important regulators of
fungiphagous nematode densities.

In a study in southeastern New Mexico, Elkins (1980) found
a similar pattern of nonstylet to stylet-bearing free-living
nematodes in buried shinnery oak (Quercus harvardii) litter.
The ratio of nonstylet to stylet bearers decreased from 4.9:1
in bags retrieved after 30 d, to 2:1 in bags retrieved after
90 d, to 1.1:1 in bags retrieved after 6 months and to 0.8:1
in bags retrieved after 1 year. Since Elkins used an overlapping sequence of bag placement and retrieval, these changes
in structure of the nematode community are a function of the
stage of decomposition of the substrate and not a seasonal

effect. For example, a set of bags retrieved after 30 d in May had a nonstylet to stylet bearing nematode ratio of 4.9:1 while a 90 d set retrieved at the same time had a ratio of 2:1.

Elkins' (1980) study also showed a seasonal increase in density of free-living nematodes in mid-summer which appeared to be independent of the stage of decomposition. However, soil moisture during June and July was 10% by weight in comparison to 5% or less in the spring months. Thus, the seasonal density increase probably occurred because of more favorable soil moisture and temperature which allowed nematode reproduction as well as migration.

Elkins' (1980) data supports our contention based on our Lepidium study that bacteriophagous nematodes are important in early stages of buried litter decomposition in the Chihuahuan desert and stylet-bearing fungiphagous nematodes are important in subsequent stages.

In decomposition of buried litter the role of nematodes is time dependent or at least dependent upon stage of decomposition (Fig. 2). In initial stages of decomposition bacteriophagous nematodes play an important role but in later stages decomposition is primarily a function of fungi and fungiphagous nematodes (Fig. 2).

Decomposition of surface litter is more complex because of the harshness of the environment. Litter temperatures on the soil surface exceed 50°C for several hours at mid-day and litter moisture drops to nearly zero and remains there during the day. Under such conditions, biological activity is not expected. However, in dense accumulations of litter (20-30 g/100 cm^2) surface litter disappears even under dry conditions (Whitford et al. 1980). This appears to be a function of the activity of soil microarthropods. We had originally hypothe-

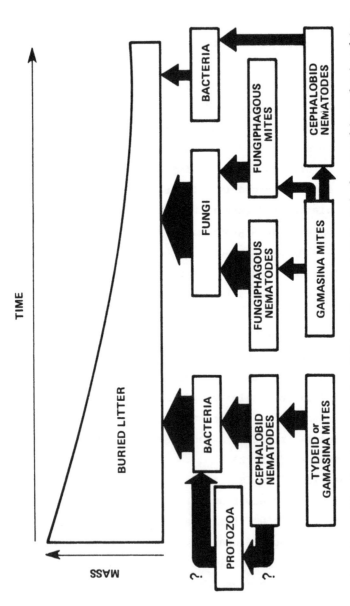

Fig. 2. Generalized scheme showing litter disappearance and the trophic relationships of the soil biota in buried leaf litter in North American desert ecosystems. The widths of the arrows indicate relative magnitude of flows between groups.

sized that the entire soil biota would be active for brief
periods every day, because immediately prior to and after
sunrise, soil surface and surface litter temperatures are low
and moisture contents elevated. We found that microarthropods
move into surface litter accumulations during this period, but
nematodes exhibited neither movement nor activity. However,
following rainfall and sufficient soil wetting, the entire
biota becomes active. Nematodes living in the litter, an
environment rich in organic matter, fungi and bacteria
appear to be active for only a short period after a rain
because within 48 h after such an event, most of the litter
nematodes are anhydrobiotic (Fig. 3). However, nematodes in

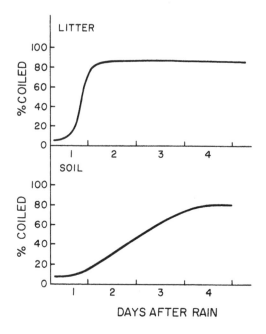

Fig. 3. Changes in proportion of free-living nematodes in
the anhydrobiotic state (percent coiled) in litter and soil
following a rain of approximately 25 mm in a Chihuahuan desert
ecosystem.

the soil are in a moist environment and remain active for a
much longer period of time (Fig. 3). We hypothesize that
fecal material produced by mites feeding in the litter during
the extended dry periods provides the energy base for soil
fungi and bacteria which are fed upon by free-living soil
nematodes (Fig. 4).

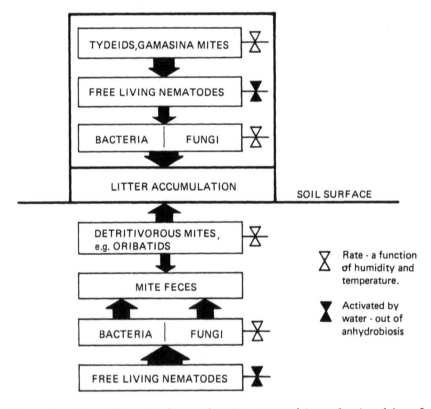

Fig. 4. Generalized scheme showing a trophic relationship of
soil and surface litter biota in a desert ecosystem. Trophic
relationships hypothesized from behavior studies of buried
litter and of soil and litter mites and nematodes in simulated
rainfall experiments. The widths of the arrows indicate
relative magnitudes of flows between groups.

Thus, the role of nematodes in decomposition of surface litter is dependent on rainfall, soil moisture and activities of detritivorous microarthropods. With the limited data currently available, it is not possible to elaborate on these hypothesized relationships. Indeed, these relationships may also pertain to other ecosystems, but until decomposition is studied in relation to the entire soil flora and fauna in other ecosystems, such generalizations must remain undocumented. The hypothetical relationships of nematodes and other soil flora and fauna in decomposition of surface litter in desert ecosystems is shown in Fig. 4. This scheme emphasizes the nematode activity as a "pulse" phenomenon triggered by rainfall. Since surface litter accumulations are rich in energy, growth of microorganisms should be very rapid after wetting. This rapidly growing microflora should provide a large readily obtainable food supply for free-living nematodes. We presume that grazing is intense in such a system. As shown in Fig. 3, however, this activity is brief because surface litter rapidly dries and the nematodes become anhydrobiotic. Although nematodes in the soil can be active over a much longer period of time, their prey (bacteria and fungi) are dispersed in three dimensional space with concentrations around the fecal deposits of the detritivorous microarthropods. Such dispersion requires considerable movement by the nematodes and much less intense feeding than is possible in litter. Thus, while the soil microclimate may be favorable for extended activity by nematodes, the soil nematodes are probably substrate limited and may naturally process less energy and nutrients than the nematodes living within the litter accumulations.

The implications of these studies go beyond decomposition and nutrient cycling. Santos (1979) observed tydeid mites

eating nematode eggs and nematodes in laboratory cultures.
Many species of mites from several orders have been shown to
be nematode predators or potential nematode predators (Muraoka
and Ishibashi 1976). Many of the prostigmatid and mesostigmatid mites found in the rhizosphere of desert grasses and
ephemerals belong to groups which feed on nematode eggs and
nematodes. Many of these acari are adversely affected by
insecticides. Reductions in density of nematode predaceous
microarthropods could, as shown by our studies, result in
large increases in nematode densities including plant parasites. Plant parasitic nematodes which lay their eggs outside
root tissues would be favored when the nematode egg predators
were reduced. In agricultural soils, tilled soils, plant
parasitic nematodes are major components of the nematode
community. There are no studies of which we are aware which
address the relationships between microarthropods and nematodes in the tilled field environment. It is possible that
agricultural practices employing large scale chemical insecticide use may result in the exchange of an insect problem for
a nematode problem. We feel this potential deserves careful
investigation.

ACKNOWLEDGMENTS

These studies were supported by NSF Grant DEB 77-16633 and
Contract CPD/B-9762 WIPP program with Sandia Laboratories to
W. G. Whitford and NSF Grant DEB 78-3445 to D. W. Freckman.

LITERATURE CITED

Babiuck, L. A., and E. H. Paul. 1970. The use of fluorescein
    isothiocyanate in the determination of the bacterial bio-
    mass of a grassland soil. Canadian Journal of Microbiology
    16:56-62.

Brock, T. D. 1971. Microbial growth rates in nature.
Bacteriological Review 35:39-58.

Chew, R. M. 1974. Consumers as regulators of ecosystems: an
alternative to energetics. Ohio Journal of Science
72:359-370.

Elkins, N. Z. 1980. Predaceous microarthropods as potential
regulators of microbial grazing during litter decomposition
in a Chihuahuan desert ecosystem. M.S. Thesis. New Mexico
State University, Las Cruces, New Mexico, USA.

Freckman, D. W., and R. Mankau. 1977. Distribution and
trophic structure of nematodes in desert soils. Ecological
Bulletin (Stockholm) 25:511-514.

Hanlon, R. D. G., and J. M. Anderson. 1979. The effects of
collembola grazing on microbial activity in decomposing
leaf litter. Oecologia 38:93-99.

Muraoka, M., and N. Ishibashi. 1976. Nematode feeding
behavior. Japanese Journal of Applied Entomology and
Zoology 11:1-7.

Nicholas, W. L. 1975. The biology of free-living nematodes.
Clarendon Press, Oxford, England.

Olson, F. C. W. 1950. Quantitative estimates of filamentous
algae. Transactions of the American Microscopical Society
59:272-279.

Santos, P. F. 1979. The role of microarthropods and nematodes
in litter decomposition in a Chihuahuan desert ecosystem.
Ph.D. Dissertation, New Mexico State University, Las Cruces,
New Mexico, USA.

Santos, P. F., J. Phillips, and W. G. Whitford. 1981. The
role of mites and nematodes in early stages of buried
litter decomposition in a desert. Ecology 62:664-669.

Santos, P. F., and W. G. Whitford. 1981. The effects of

microarthropods in litter decomposition in a Chihuahuan
desert ecosystem. Ecology 62:654-663.

Sohlenius, B. 1980. Abundance, biomass, and contribution
to energy flow by soil nematodes in terrestrial ecosystems. Oikos 34:186-194.

Whitford, W. G., M. Bryant, G. Ettershank, J. Ettershank, and
P. F. Santos. 1980. Surface litter breakdown in a
Chihuahuan desert ecosystem. Pedobiologia 4:243-245.

Yeates, G. W. 1973. Nematoda of a Danish beech forest. II.
Production estimates. Oikos 24:179-185.

THE ROLE OF BACTERIVOROUS AND FUNGIVOROUS
NEMATODES IN CELLULOSE AND CHITIN DECOMPOSITION IN
THE CONTEXT OF A ROOT/RHIZOSPHERE/SOIL CONCEPTUAL MODEL

J. A. Trofymow and D. C. Coleman

INTRODUCTION

Previous papers in this symposium have noted that nematodes
are ubiquitous, existing in an array of terrestrial and aquatic
habitats. Yeates and Coleman (this volume) reviewed the diversity of reproductive and feeding strategies that allow nematodes to utilize virtually every sort of substrate available
to metazoa.

To better understand specific aspects of the role of nematodes in decomposition, workers have conducted a series of
experiments in laboratories around the world of grassland
(Colorado, Coleman et al. 1977, 1978), pine forest (Sweden,
Bååth et al. 1978, in press), and sewage sludge-amended soils
(Syracuse, New York, Abrams and Mitchell 1980). We will concentrate our report on a conceptual model of nematodes in a
root/rhizosphere/soil system and on recent results from model
soil microcosm systems.

Carbon Inputs and Decomposition in Grassland Soils

The soil environment can be viewed as an interconnected network
of pores, varying in size and water content, through which
nutrients and organisms can migrate. Maximum microfloral
activity occurs at sites of monomeric and oligomeric forms of
reduced carbon (especially plant material). These sites may be
close to the plant, as in the rhizosphere, where root exudate
or sloughed root material is present, or in senescent and dead
roots (a more resistant carbon source). Other sites of reduced carbon are at heterogeneously distributed clumps of

organic matter entering the soil as litter. In the shortgrass prairie about 23% of the organic carbon enters the soil as aboveground litter, 54% as detrital roots and 23% as root exfoliates and exudates (Coleman 1976).

Coleman et al. (in press) postulate the existence of pathways of catabolism, involving labile and nonlabile substrates, rather than organism-oriented pathways. In a series of experiments Woods et al. (in press), Anderson et al. (1978), Coleman et al. (1978a,b), Cole et al. (1978), and Herzberg et al. (1978) found that combining simulated root exudate and bacteria, amoebae, and nematode grazers resulted in the marked remineralization of P and, in some instances, N, over periods of several weeks.

We extended those studies to model nonlabile substrates (chitin and cellulose), the chitinoclastic and cellulolytic microflora (bacteria and fungi), and the bacterivorous and fungivorous nematode grazers that feed upon them.

We postulated that decomposition would increase during grazing and that a significant synergism (i.e., enhanced total decomposition in the presence of two substrates or two or more species of primary decomposers) would occur, as was noted by Dommergues and Mangenot (1970).

During the experiments, we looked for the aspects of functional group dynamics that one might expect because of the diverse nature of feeding. Thus, our bacteria feeders are holophagic, digesting only about 50-80% of the food ingested (Smerda et al. 1971); whereas hyphal-feeders, have a piercing stylet, suck cell sap (cytoplasm) out of the hyphae, as do plant root feeders. The feeding mode may have a profound effect on overall metabolic activity and nutrient remineralization.

A REPRESENTATION OF A ROOT/RHIZOSPHERE/SOIL SYSTEM

Fig. 1 [representing roots in a shortgrass prairie in northeastern Colorado dominated by Bouteloua gracilis [H.B.K.
(Griffiths)] (blue grama)] is a conceptual diagram of the important interactions occurring in a root/rhizosphere/soil
system. The following description in conjunction with the
center section of Figure 1 provides a simple illustration of
the importance of different root regions.

Root/Rhizosphere

Major phenological changes occur during the growth and extension of a root. Mucigel, the zone of elongation, and root
exudation are associated with the growing root tip, an area of
rapid growth, sloughing of cells, and exudation of low molecular weight carbon compounds (Rovira et al. 1979). Mucigel
produced at the root tip may function in several ways: as a
lubricant enhancing root penetration through the soil; as a
medium through which extracellular enzymes of root or microbial origin diffuse to come in contact with clay-fixed or
particulate sources of substrate; and as a source of substrates
for growing microflora, which may protect the root by screening
it from plant pathogens (Atkinson et al. 1972).

A root hair zone develops behind the root tip and could enhance nutrient uptake either due to the increased surface area
of the root hairs (Brady 1974) and/or by anchoring the root to
maintain close contact with the soil (Newman 1974).

Within the root hair zone and as the root ages, mycorrhiza
(endomycorrhizae in the diagram) are formed with particular
species of symbiotic fungi, well known to enhance P (Rhodes
and Gerdemann 1980) and perhaps N uptake (Pang and Paul 1980)
for the plant in exchange for reduced carbon from the plant.

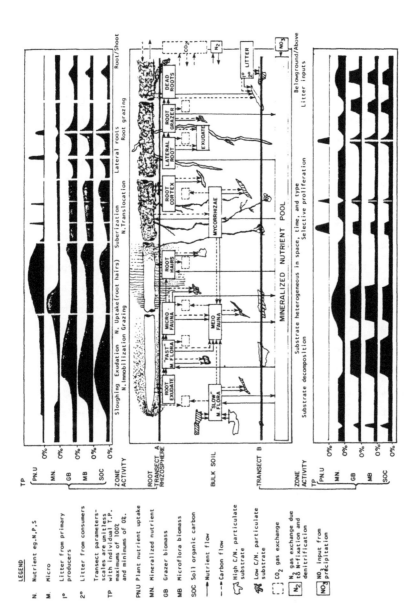

Fig. 1. Conceptual diagram of a root/rhizosphere/soil system. Refer to text for explanation.

As the root becomes suberized, lignin and cellulose content increases (Coleman 1976), mineral nutrient uptake ability decreases, and cortical cells are sloughed, releasing nonlabile forms of carbon. Development of new lateral roots renews the nutrient uptake capabilities of the root and increases the level of labile carbon in the soil (Coleman 1976). Death of roots provides an input of resistant, high C to N ratio (C:N) substrate.

The following is a brief description of how root exudate might enhance nutrient availability to the plant. Reference should be made to the transect through the rhizosphere (Transect A, Fig. 1) parallel to a growing root and to the idealized relative levels of soil organic carbon (SOC), microfloral biomass (MB), grazer biomass (GB), mineralized nutrient (MN), and rate of plant nutrient uptake (PNU) in the expanded transect (upper graphs, Fig. 1). High C:N primarily monomeric material produced at the root tip (Fig. 1, SOC) is readily metabolized by microbes. The populations rapidly increase, immobilizing any (Fig. 1, MN) nutrient in or diffusing to the root region and any available organic form of the nutrient. The increase in microfloral biomass (Fig. 1, MB) stimulates the microfauna (Fig. 1, GB) populations which begin to mineralize nutrients (Coleman et al. 1978a). As long as sufficient carbon is available, nutrients will remain immobilized, primarily in microfloral biomass. However, the root extends and matures, the zone of peak exudation moves away with the extending root, and the microhabitat becomes drier as water is taken up by the plant. The microfauna, mineralizing nutrients, continue to graze on the declining populations of microflora then either encyst or migrate. Nutrients are mineralized (Fig. 1, MN, beginning of root hair zone) when the root hairs begin to develop, thereby increasing the pool of nutrients available for plant uptake.

Death of the root hairs (Fig. 1; end of root hair zone)
provides a pulse of easily assimilated carbon, briefly increasing microflora and -fauna populations. Because they are
less labile, cortical cells (Fig. 1, root cortex) sloughed
during suberization will support a lower, slower-growing
microbe population. Organisms capable of decomposing such
materials as lignin and cellulose will predominate where those
substrates are the major components of the organic matter
(Alexander 1977). Young lateral roots would have similar
effects to those described for young primary roots.

Root grazers (e.g., ectoparasitic nematodes), occurring
primarily with young unsuberized roots, add to the pool of
plant-derived soil carbon (Fig. 1, root grazer). Excretory
products, "leaks" from the root caused by injury to the root,
and dead roots all add to the available pool of organic soil
carbon.

Bulk Soil

A major portion of the soil organic carbon occurs away from the
root in the bulk soil and is in a more resistant particulate
form. An idealized transect through this area (Fig. 1, Transect B and its expansion) would show microflora associated
primarily with the "islands" of organic matter. Depending on
the origin, C:N, and history of the organic matter, nutrients
at these sites would be either immobilized or mineralized.
Sites of particulate organic matter supporting large microfloral populations could attract and support micro- and meio-
faunal populations.

Mycorrhizal fungi may selectively proliferate (Fig. 1)
where a nutrient is being mineralized, and bacterial activity
may increase next to hyphae because of hyphal exfoliates or
other material from hyphal walls.

The box and arrow diagram in Fig. 1 summarizes some of the major flows of carbon and nutrients between the biotic and mineral components of a root/rhizosphere/soil system. The meiofauna and the mycorrhizal fungi can be viewed as "shunts" between the labile and nonlabile pathways of catabolism. Meiofauna disperse primary decomposers (Powell 1971) and translocate nutrients and carbon as biomass. The mycorrhizal fungi translocate nutrients from the bulk soil to the root; and, when the hyphae die, their biomass (of plant-derived carbon) becomes part of the bulk soil.

DECOMPOSITION EXPERIMENTS

A review of a few of our nonlabile substrate decomposition experiments at Colorado State University follows:

Substrates

Because a large part of the vegetation added to soil is cellulosic, the decomposition of this carbohydrate has special significance in carbon cycling. Cellulose consists of long, linear chains of glucose units bound together in β-linkages at carbon atoms 1 and 4. Typically, 2,000 to 10,000 glucoside residues are linked in one chain (Sihtola and Neimo 1975). Elementary chains do not typically exist in native cellulose but tend to form fibrils, in which polymer chains are oriented in parallel and are firmly bound together by a large number of strong hydrogen bonds. The fibrils are arranged in higher-order bundles, which may be structurally interlinked with lignin in plant cell walls.

The cellulose used in these experiments was a purified microcrystalline cellulose with particle sizes of 5-20 μm (Whatman #CC41 Cellulose Powder).

Chitin is a structural compound in the cell walls of filamentous fungi, egg shells of nematodes, and the exoskeleton of
arthropods (Muzzarrelli 1977). Because chitin is high in nitrogen (C:N$\cong$6·6) and much of it is formed by microbial biosynthesis,
it is undoubtedly an important substance in the carbon cycle in
the soil. Chitin is a long, linear chain of N-acetylglucos-
amine units, a polymer similar to cellulose, except that one of
the hydroxyl groups of each glucoside residue is replaced by an
acetyl amino group. Individual chains can be bound, in parallel or antiparallel, into microfibrils of varying crystalline
states and hence decomposability. Arthropod cuticle is a
composite material with microfibrils of α-chitin (mostly crystalline) randomly embedded in a protein matrix.

The chitin used in these experiments was from a crustacean
cuticle that was washed in an acid-alkaline solution and ball-
milled (described in Gould et al., in press).

Organism Isolations

Chitin- and cellulose-decomposing bacteria, fungi, and actinomycetes were isolated from soils of the Pawnee National Grasslands, a shortgrass prairie in northeastern Colorado dominated
by Bouteloua gracilis. The decomposers were isolated by en-
riching the soil with cellulose or chitin or by a buried slide
technique (Parkinson et al. 1971), then tested for their relative abilities to decompose chitin (NACG production, Reissing
et al. 1955) and cellulose (Agar-diffusion assay, Tansey 1971).

Palatibility studies were conducted using the bacterial iso-
lates and a Pelodera sp. isolated from the chitin enrichment.
Nematodes were sterilized by passing an adult female through
alternate series of wells containing dilute (1/2 strength)
nutrient broth with or without an antibiotic mixture of strep-
tomycin sulfate, penicillin, and fungizone. Aphelenchus avenae

cultures on <u>Rhizoctonia</u> <u>solani</u>, obtained from D. Freckman,
University of California at Riverside, were cleaned and transferred to a <u>Fusarium</u> <u>oxysporum</u> isolated from a soil enrichment.
<u>A. avenae</u> occurs abundantly in the native soil (Smolik, unpub-
lished data).

Microcosm Design and Sampling

The organisms were inoculated into 50-ml Erlenmeyer flasks
containing 20 g of soil from the Renohill-Shingle complex, a
sandy loam from the Pawnee Grasslands. The soil was moistened
and autoclaved twice at 24-h intervals, amended with cellulose
or chitin, or both, autoclaved a third time, and dried. One
ml of a washed liquid culture of bacteria (<u>Flavobacterium</u> sp.)
or fungi (<u>F</u>. <u>oxysporum</u>) was inoculated into the microcosms.
The soil was brought to 15% moisture w/v (field capacity) with
a mineral salts solution (RSS, Herzberg et al. 1978) with or
without an ammonium sulfate nitrogen amendment. The microcosms
were sealed in 500-ml Mason jars containing 1.8 ml of 1 N NaOH
to absorb CO_2. The frequency of the titrations of the NaOH
was initially every 24 h but was decreased as the level of
respiration decreased. Nematode grazers were inoculated into
the systems 2-4 d after the microflora were inoculated.

Three replicate microcosms were destructively sampled at the
completion of each experiment. In some experiments, additional
microcosms were prepared to allow sampling during the experiment. Analyses were made of NH_4^+-N (modified Conway, Stanford
et al. 1973), inorganic P (bicarbonate extractable, Olsen et al.
1954), bacteria populations (soil dilutions plated on nutrient
agar), gravimetric soil moisture, nematode numbers (modified
Baerman, Anderson and Coleman 1977) and, in some cases, chitin
remaining in the soil (alkaline distillation, Bremner and Shaw
1954).

Chitin Decomposition

Chitin was added to the microcosms at a level of 3000 µg chitin-$C \cdot g^{-1}$ dry soil. Microcosms were destructively sampled on days 24, 37, 80. By day 80 significantly more chitin was decomposed and more NH_4^+-N mineralized in the _Fusarium_ treatment than in either the grazed or ungrazed bacteria treatments (Fig. 2). Nematode-grazed bacteria (_Flavobacterium_ sp.) mineralized a significant amount of N from native organic-N sources (Fig. 2). Bacterial populations were initially higher in the grazed treatments but decreased as the nematode population increased (Fig. 3). Bacterial biomass may have increased in the presence of nematodes because of the bacteria's utilization of nematode excretory and fecal products, particularly amino acids (Anderson et al., submitted). In addition, the nematodes could disperse bacteria (Anderson et al., in press) adsorbed on the nematode's cuticle (Powell 1971). Increased metabolic activity in the presence of a grazer has been observed several times (Coleman et al. 1977, Anderson et al. 1978, Coleman et al. 1978a, Elliott et al. 1980) and is attributed to the grazer's keeping the bacteria population in more active growth phases.

Cellulose Decomposition

Cellulose additions of 800 µg cellulose-$C \cdot g^{-1}$ dry soil were made to all microcosms, half of which received additions of 115 µg $(NH_4)_2SO_4$-$N \cdot g^{-1}$ dry soil. Addition of NH_4^+ increased the level of C mineralization by the bacteria (Fig. 4) to the level in microcosms with nematode-grazed bacteria. By the end of the experiment the unamended nematode-grazed bacterial treatment had mineralized significantly more of the native organic-N (Table 1) and had higher numbers of bacteria than the bacteria-alone treatments. Carbon mineralization in the N-amended

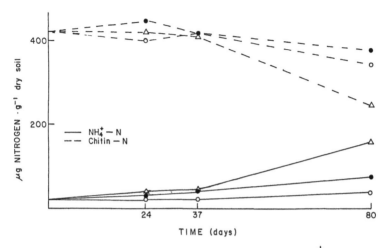

Fig. 2. Mean levels of chitin (---) and (——) NH_4^+-N in the chitin decomposition experiment. $\triangle$ = Fungus, O = Ungrazed bacteria, ● = Nematode grazed bacteria treatment.

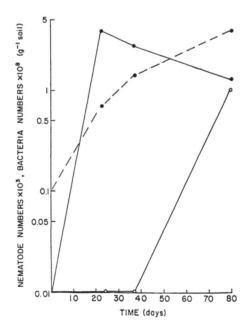

Fig. 3. Mean bacteria (——) and nematode (---) numbers in the chitin decomposition experiment. O = Ungrazed bacteria, ● = Nematode grazed bacteria.

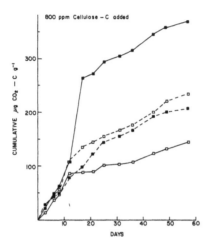

Fig. 4.   Mean cumulative CO_2 evolution from bacterial treatments in the cellulose decomposition experiment with (---) and without (——) the addition of 115 μg $(NH_4)_2SO_4$-N·g^{-1}. $\square$ = Ungrazed bacteria, $\blacksquare$ = Nematode grazed bacteria.

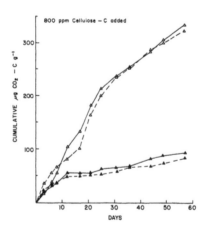

Fig. 5.   Mean cumulative CO_2 evolution from fungal treatments in the cellulose decomposition experiment with (---) and without (——) the addition of 115 μg $(NH_4)_2SO_4$-N·g^{-1}. $\triangle$ = Ungrazed fungus, $\blacktriangle$ = Nematode grazed fungus.

TABLE 1.   Decomposition of Cellulose in Microcosms With and
Without Nitrogen Amendments.   Treatments Include Microflora
Alone and in Combination With Adult or Juvenile Microbivorous
Nematodes.   Values are Means of Three Replicates.

Treatment	Counts of Organisms	NH_4^+-N $(\mu g/g)$	Pi $(\mu g/g)$
No nitrogen added			
Bacteria (Flavobacterium sp.)	$76 \cdot 10^5 \cdot g^{-1}$	22	19
Bacteria Pelodera sp.	$838 \cdot 10^5 \cdot g^{-1}$		19
" "	$108 \text{ Adults} \cdot g^{-1}$	34	19
" "	$370 \text{ J}_2\text{-J}_4 \cdot g^{-1}$		
Fungus (Fusarium oxysporum)	--	32	18
Fungus with A. avenae	--		
" "	$9 \cdot g^{-1}$	24	18
Uninoculated control	--	24	18
115 ppm ammonium sulfate-N			
Bacteria	$306 \cdot 10^5 \cdot g^{-1}$	133	17
Bacteria with Pelodera sp.	$434 \cdot 10^5 \cdot g^{-1}$		17
" "	$36 \text{ Adults} \cdot g^{-1}$	140	17
" "	$189 \text{ J}_2\text{-J}_4 \cdot g^{-1}$		
Fungus	--	159	19
Fungus with A. avenae	--		
" "	$4 \cdot g^{-1}$	136	19
Uninoculated control	--	139	17

grazed system was not significantly different from that in the
ungrazed systems perhaps because of the toxic effects of am-
monia on the nematodes, reflected, perhaps, in lowered nematode
numbers.   The amount of C mineralized by the fungi was un-
affected by the N amendment (Fig. 5).   In contrast to the bac-
teria-grazed and -ungrazed systems, significantly less C was
mineralized by the nematode-grazed hyphae than by the fungi
alone.   Large nematode inoculations (25 individuals$\cdot g^{-1}$),
combined with a short initial incubation period for fungi (48
h), may have resulted in the fungi's being overgrazed.   The
fungi mineralized a significant amount of native organic-N
(Table 1).

Decomposition of Cellulose and Chitin

Cellulose (314 µg C·g^{-1} dry soil) was added to all microcosms; some microcosms received an additional amendment of 686 µg chitin-C·g^{-1} dry soil (total C = 1000 µg; C:N = 10:1). Throughout the experiment significantly more N was mineralized in the nematode-grazed bacteria treatment than in the ungrazed treatment (Fig. 6). As was seen in the previous experiments, nematode grazed bacterial populations (Fig. 7) were initially higher than ungrazed. A bacterial contaminant unable to decompose cellulose or chitin was introduced with the fungus grazer A. avenae. Presumably, the bacteria could compete with the fungi for limiting nutrients or for breakdown products produced by fungal chitinase. Even with substantial bacteria populations (Fig. 7) the grazed-fungus treatment mineralized more N than the ungrazed treatment (Fig. 8). The nematode-grazed bacteria mineralized more N than either of the fungus treatments.

In the cellulose alone-amended microcosms about 300 µg of CO_2-C·g^{-1} dry soil was respired in all biological treatments except the bacteria-alone treatment (Fig. 9). Microcosms containing the most complex biota had higher initial rates of CO_2 evolution but the rate decreased earlier than in microcosms containing the fungal and grazed bacteria treatments. Microcosms with a cellulose and chitin amendment evolved more CO_2 (Fig. 10) than the same biological treatment with cellulose alone (Fig. 9), except for the ungrazed bacteria treatment, where no significant difference in CO_2 evolution occurred. The treatment with the most complex biotic treatment mineralized significantly more C than any of the other treatments. In both C amendments the nematode-grazed bacteria respired more than the ungrazed bacteria.

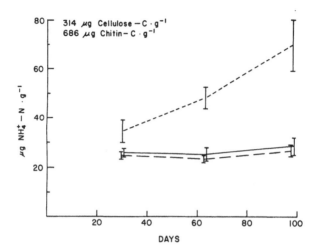

Fig. 6. Mean soil NH₄-N levels with 95% confidence intervals in bacterial treatments of the cellulose and chitin decomposition experiment. ——— = Ungrazed bacteria, --- = Nematode grazed bacteria, — — — = Uninoculated control.

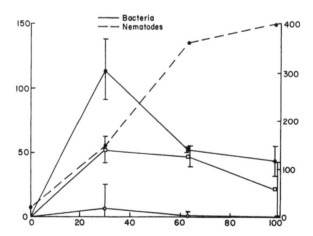

Fig. 7. Mean bacteria (———) and nematode (---) numbers with 95% confidence intervals for treatments in the cellulose and chitin decomposition experiment. O = Ungrazed bacteria, ● = Nematode grazed bacteria, □ = Contaminating bacteria in the nematode grazed fungus treatment.

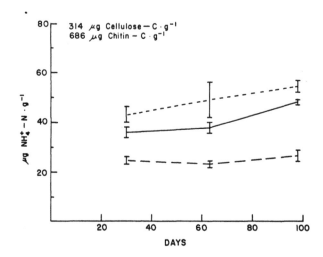

Fig. 8. Mean soil NH_4^+-N levels with 95% confidence intervals in fungal treatments of the cellulose and chitin decomposition experiment. ——— = Ungrazed fungus, --- = Nematode grazed fungus, — — — = Uninoculated control.

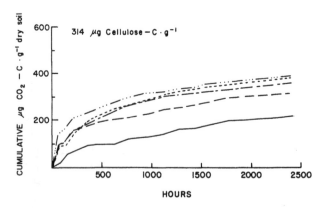

Fig. 9. Mean cumulative CO_2 evolution from microcosms containing 314 µg cellulose-C·g^{-1} amendment in the cellulose and chitin decomposition experiment. ——— = Ungrazed bacteria, --- = Nematode grazed bacteria, — — = Ungrazed fungus, —·— = Nematode grazed fungus, —···— = Nematode grazed bacteria with nematode grazed fungus.

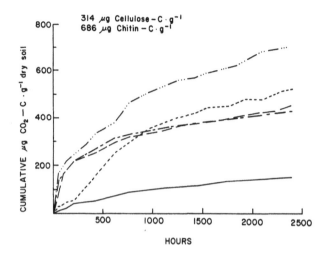

Fig. 10. Mean cumulative CO_2 evolution from microcosms containing 314 µg cellulose-C·g^{-1} and 686 µg chitin-C·g^{-1} amendment in the cellulose and chitin decomposition experiment.
——————— = Ungrazed bacteria, --- = Nematode grazed bacteria,
—— —— = Ungrazed fungus,—— -—— = Nematode grazed fungus,
——···—— = Nematode grazed bacteria with nematode grazed
fungus.

   In summary, in some cases nematode grazers can increase the
rate of organic matter decomposition. This could occur by
maintaining microfloral populations in a log phase growth, by
mineralizing nutrients that would otherwise remain immobilized
in stationary-phase microflora, or by dispersing relatively
immobile microflora. In other instances, if microbial populations (particularly fungi) are very heavily grazed, system-level
activity may be significantly inhibited.

ACKNOWLEDGMENTS

This research was supported by NSF Research Grants DEB 78-11201
and 80-04193. Thanks are extended to C. Cambardella, C. R.
Morley, R. Hays, and J. S. Frey for technical assistance. We
also thank E. T. Elliott, W. D. Gould, R. V. Anderson, C. R.
Morley and T. V. St. John for their contributions towards the
conceptual model.

LITERATURE CITED

Abrams, B. I., and M. J. Mitchell. 1980. Role of nematode-
    bacterial interactions in heterotrophic systems with em-
    phasis on sewage sludge decomposition. Oikos 35:404-410.
Alexander, M. 1977. Introduction to soil microbiology. 2nd
    Edition. John Wiley and Sons, Inc., New York, New York,
    USA.
Anderson, R. V., and D. C. Coleman. 1977. The use of glass
    microbeads to simulate the natural environment for ecolog-
    ical experiments with bacteriophagic nematodes. Journal
    of Nematology 9:319-322.
Anderson, R. V., E. T. Elliott, J. F. McClellan, D. C. Coleman,
    C. V. Cole, and H. W. Hunt. 1978. Trophic interactions
    in soils as they affect energy and nutrient dynamics. III.
    Biotic interactions of bacteria, amoebae, and nematodes.
    Microbial Ecology 4:361-371.
Anderson, R. V., J. A. Trofymow, D. C. Coleman, C. P. Reid.
    Phosphorus mineralization in spent oil shale as affected
    by nematodes. Soil Biology and Biochemistry (in press).
Anderson, R. V., W. D. Gould, L. E. Woods, C. Cambardella,
    R. E. Ingham, and D. C. Coleman. Organic and inorganic
    nitrogenous losses by nematodes in soil. Oikos (submitted).
Atkinson, T. G., J. L. Neal, Jr., and R. I. Larson. 1972.

Genetic control of the rhizosphere microflora of wheat.
Annual Review of Phytopathology 10:116-122.

Bååth, E., U. Lohm, B. Lundgren, T. Rosswall, B. Söderström,
B. Sohlenius, and A. Wiren. 1978. The effect of nitrogen
and carbon supply on the development of soil organism
populations and pine seedlings: A microcosm study. Oikos
31:153-163.

Bååth, E., U. Lohm, B. Lundgren, T. Rosswall, B. Söderström,
and B. Sohlenius. Impact of microbial-feeding animals on
the total soil activity and nitrogen dynamics: A soil
microcosm experiment. Oikos (in press).

Brady, N. C. 1974. The nature and properties of soils.
MacMillan Publishing Company, Inc., New York, New York, USA.

Bremner, J. M., and H. Shaw. 1954. Studies on the estimation
and decomposition of amino sugars in soil. Journal of
Agricultural Science 44:152-159.

Cole, C. V., E. T. Elliott, H. W. Hunt, D. C. Coleman, and
M. K. Campion. 1978. Trophic interactions in soils as
they affect energy and nutrient dynamics. V. Phosphorus
transformations. Microbial Ecology 4:381-387.

Coleman, D. C. 1976. A review of root production processes
and their influence on soil biota in terrestrial ecosystems. Pages 417-434 in J. M. Anderson and A. Macfadyen,
editors. The role of terrestrial and aquatic organisms in
decomposition processes. Blackwell Science Publications,
Oxford, England.

Coleman, D. C., C. V. Cole, R. V. Anderson, M. Blaha, M. K.
Campion, M. Clarhold, E. T. Elliott, H. W. Hunt, B.
Schaefer, and J. Sinclair. 1977. Analysis of rhizosphere-
saprophage interactions in terrestrial ecosystems. in
U. Lohm and T. Persson, editors. Soil organisms as com-

ponents of ecosystems. Ecological Bulletin (Stockholm)
25:299-309.

Coleman, D. C., R. V. Anderson, C. V. Cole, E. T. Elliott, L.
Woods, and M. K. Campion. 1978a. Trophic interactions in
soils as they affect energy and nutrient dynamics. IV.
Flows of metabolic and biomass carbon. Microbial Ecology
4:373-380.

Coleman, D. C., C. V. Cole, H. W. Hunt, and D. A. Klein. 1978b.
Trophic interactions in soils as they affect energy and
nutrient dynamics. I. Introduction. Microbial Ecology
4:345-349.

Coleman, D. C., C. P. Reid, and C. V. Cole. Biological
strategies of nutrient cycling in soil systems. Advances
in Ecological Research 13 (in press).

Dommergues, Y. R., and F. Mangenot. 1970. Ecologie microbienne
du sol. Masson et Cie., Paris, France.

Elliott, E. T., R. V. Anderson, D. C. Coleman, and C. V. Cole.
1980. Habitable pore space and microbial trophic interactions. Oikos 35:327-335.

Gould, W. D., R. J. Bryant, J. A. Trofymow, R. V. Anderson,
E. T. Elliott, and D. C. Coleman. 1980. Chitin decomposition in a model soil system. Soil Biology and Biochemistry
(in press).

Herzberg, M. A., D. A. Klein, and D. C. Coleman. 1978. Trophic
interactions in soils as they affect energy and nutrient
dynamics. II. Physiological responses of rhizosphere and
non-rhizosphere bacteria. Microbial Ecology 4:351-359.

Muzzarrelli, R. A. A. 1977. Chitin. Pergamon Press, New York,
New York, USA.

Newman, E. I. 1974. Root and soil water relations in the
plant root and its environment. Page 363 in E. W. Carson,
editor. The plant root and its environment. The University

Press of Virginia, Charlottesville, Virginia, USA.

Olsen, S. R., C. V. Cole, W. S. Watanabe, and L. A. Dean. 1954. Estimation of available phosphorus in soils by extraction with sodium bicarbonate. United States Department of Agriculture, Circular 939, Washington, District of Columbia, USA.

Pang, P. C., and E. A. Paul. 1980. Effects of vesicular-arbuscular mycorrhiza on ^{14}C and ^{15}N distribution in nodulated fababeans. Canadian Journal of Soil Science 60:241-250.

Parkinson, D., T. R. G. Gray, and S. T. Williams. 1971. Methods for studying the ecology of soil micro-organisms. IBP Handbook 19. Blackwell Scientific Publications, Oxford, England.

Powell, N. T. 1971. Interactions of plant parasitic nematodes with other disease-causing agents. Pages 119-135 in B. M. Zuckerman, W. F. Mai, and R. A. Rhode, editors. Plant parasitic nematodes. Volume II. Academic Press, New York, New York, USA.

Reissing, J. L., J. L. Strominger, and L. F. LeLoir. 1955. A modified colorimeter method for the determination of N-acetylamino sugars. Journal of Biological Chemistry 217:959-966.

Rhodes, L. H., and J. W. Gerdemann. 1980. Nutrient translocation in vesicular-arbuscular mycorrhizae. Pages 150-168 in C. B. Cook, P. W. Pappas, and E. D. Rudolph, editors. Cellular interaction in symbiosis and parasitism. The Ohio State University Press, Columbus, Ohio, USA.

Rovira, A. D., R. C. Foster, and J. K. Martin. 1979. Note on terminology, origin, nature and nomenclature of the organic materials in the rhizosphere. Pages 1-4 in J. L. Harley and R. S. Russell, editors. The soil-root interface. Academic Press, London, England

Sihtola, H., and L. Neimo. 1975. The structure and properties of cellulose. Pages 10-25 in M. Bailey, T. M. Enari, and M. Linko, editors. Symposium on enzymatic hydrolysis of cellulose. SITRA, Helsinki, Finland.

Smerda, S. M., H. J. Jensen, and A. W. Anderson. 1971. Escape of Salmonellae from chlorination during ingestion by Pristionchus lheritieri (Nematoda:Diplogasterinae). Journal of Nematology 3:201-204.

Stanford, G., J. N. Carter, E. C. Simpson, Jr., and D. E. Schwaninger. 1973. Nitrate determination by a modified Conway microdiffusion method. Journal of American Organization of Analytical Chemists 56:1365-1368.

Tansey, M. R. 1971. Agar-diffusion assay of cellulolytic ability of thermophilic fungi. Archiv fuer Mikrobiologie 77:1-11.

Woods, L. E., C. V. Cole, E. T. Elliott, R. V. Anderson, and D. C. Coleman. Nitrogen transformations as affected by bacterial-microfaunal interactions. Soil Biology and Biochemistry (in press).

Part III
Synthesis and Validation

MODEL SYNTHESIS AND VALIDATION: PRIMARY CONSUMERS

R. McSorley, J. M. Ferris and V. R. Ferris

INTRODUCTION

In recent years, the subject of mathematical models for
primary consumers has been receiving increased attention
in the applied agricultural sciences (Ruesink 1976). Ruesink
(1975) defines a model as "an imitation and representation
of the real world." Such a definition could include not only
sophisticated computer simulations but also intuitive models
based on experience. Two types of models which pertain
directly to nematode-plant interactions are nematode population models and plant damage models.

NEMATODE POPULATION MODELS

Models simulating nematode population changes are available
for several nematode-plant associations, including Meloidogyne
arenaria (Neal) Chitwood on grapevine (Ferris 1976, 1978a),
Globodera spp. and Heterodera spp. on potato (Jones et al.
1978, Mugniery 1976), Pratylenchus penetrans (Cobb) Filipjev
and Schuurmans-Stekhoven on potato (Bird et al. 1976),
P. hexincisus Taylor and Jenkins on corn (McSorley and Ferris
1979), and P. minyus Sher and Allen on wheat (Kimpinski et al.
1976). In general, these models simulate population fluctuations during one growing season. An example is the seasonal
simulation of M. arenaria egg and larval counts by MELSIM
(Ferris 1976, 1978a). However, population changes may also be
projected over several years, as shown by Jones et al. (1978)
for potato cyst nematodes under various rotation schemes.

MODEL SYNTHESIS

Procedures used in the construction of population models
depend on the particular situation. However, some general
guidelines are applicable to most situations. Miles (1974)
has outlined the following steps to model development:

1. Identify the problem
2. Specify system boundaries
3. Formulate the model
4. Write the computer code
5. Verify the model
6. Validate the model
7. Perform a sensitivity analysis
8. Use the model

Obviously, the problem requires clear and unambiguous definition since mathematical descriptions are to be employed. The
system, or that part of the real world to be modeled, must be
defined early and provisions made to monitor external variables
which will affect the operation of the system. In modeling
biological systems weather data such as rainfall and air or
soil temperatures are usually monitored as external variables.
During these preliminary stages of system identification and
definition, assumptions about the system should be explicitly
stated.

Two general approaches have been used in formulating models
of nematode population fluctuations. In one method, models
are developed from theoretical equations describing the
dynamics of population growth (Jones et al. 1978), or the
rates at which cohorts advance through various life stages
can be described. The life-table approach has been used widely
in entomology (Ruesink 1975), and has been used successfully
by Ferris (1976) in simulating nematode population changes.
A second approach to model formulation involves the regression

analysis of field data. Kimpinski et al. (1976) described
population fluctuations of P. minyus in terms of various
environmental and biotic variables. McSorley and Ferris
(1979) used patterns in population fluctuations of P.
hexincisus to simulate nematode buildup in corn roots. In
practice, the two approaches have overlapped to some extent;
for example, Ferris (1976) used regression analyses to define
interactions among subcomponents at several points in MELSIM.
Construction of a model using the first approach is more
difficult since considerable data on the various life stages
must be gathered and all data gaps filled by preliminary
experimentation before model synthesis can begin. Regression
models often do not require such detailed data and thus can be
constructed more quickly. Since they are derived from field
data, they may reproduce field situations successfully at a
relatively early stage of testing. One important advantage
of models developed from life table data is their flexibility.
When conditions change, the model may be modified by changing
individual components rather than modifying the entire model.
A regression model may need extensive modification if there is
a change in the conditions for which it was developed.

The steps of model verification, validation, and sensitivity
analysis can be illustrated by using as an example the model
PHEX (McSorley and Ferris 1979), which simulates the increase
of P. hexincisus in corn roots during the course of a growing
season. Formulation and coding of PHEX have been discussed
elsewhere (McSorley and Ferris 1979), but major components of
PHEX are illustrated (Fig. 1). Weather and agronomic data
are external to the system and must be supplied as input.
Counts of P. hexincisus in soil prior to planting must also
be supplied for each simulation. Soil temperatures in the
root zone of the host plant are used to develop the heat

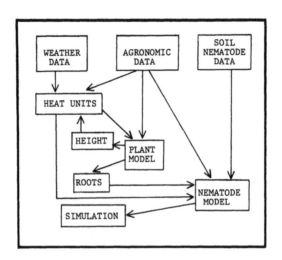

Fig. 1.   Major components of PHEX

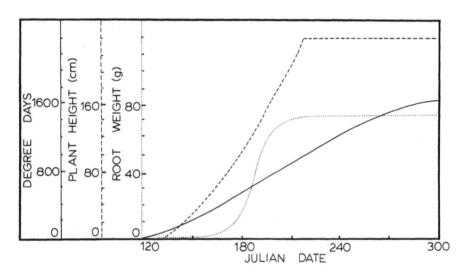

Fig. 2.   Example output of PHEX: heat units (10°C base),
plant height, and root weight.

units which drive both the plant and nematode components of
the model (McSorley and Ferris 1979). Model output (Fig. 2)
consists of simulated values of centigrade degree days accu-
mulated after planting using a 10°C base, as well as estimates
of plant height and dry root weight. In addition, a simulation of total P. hexincisus in the root system is provided
(Fig. 3). Pratylenchus hexincisus per gram of dry root weight
is also simulated, since it is a more convenient quantity for
field assay.

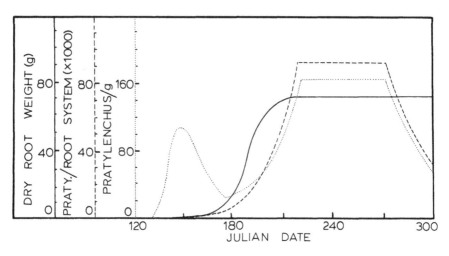

Fig. 3. Example output of PHEX: root weight, P. hexincisus
per root system, and P. hexincisus per gram of dry weight.

MODEL VERIFICATION

Fig. 4 shows the simulated values of P. hexincisus per gram
of root weight using the input parameters for a field near
Shadeland, Indiana sampled in 1975. The 95% confidence intervals around the means of actual samples collected in this
field are superimposed on the simulation and, in fact,

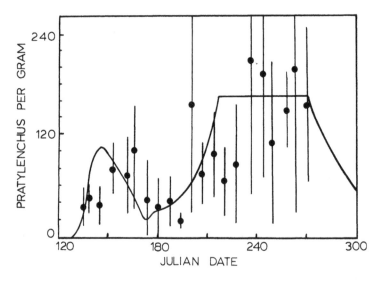

Fig. 4. Simulated counts of P. hexincisus per gram of dry root weight with means and 95% confidence intervals of actual samples superimposed.

represent a data set used in the development of PHEX (McSorley and Ferris 1979). In 17 of 21 cases, the simulated values fell into these 95% confidence intervals and, in many cases, the actual sample means and the simulated values are close. Verification has been accomplished since the model responds as expected to known inputs.

SENSITIVITY ANALYSIS

Substitution of 1976 or 1977 weather data for the 1975 data used in the trial run (Fig. 4) had little effect on the magnitude of the initial peak in the curve of P. hexincisus per gram of root weight. Three additional scenarios were examined, using once again the input data from the Shadeland field in

1975 as well as data from the same field in 1976, and from a
field near Romney, Indiana, sampled in 1976 (McSorley and
Ferris 1979).   Initial soil populations of P. hexincisus for
these sites were 102/500 cm^3, 167/500 cm^3, and 22.5/500 cm^3,
respectively.   Results from these sample runs (Fig. 5)

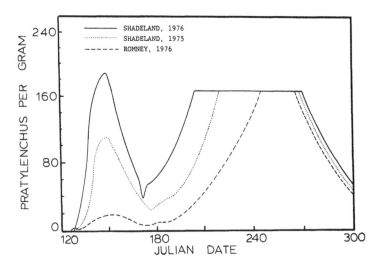

Fig. 5.   Comparison of simulated counts of P. hexincisus per
gram of dry root weight for three different initial P.
hexincisus populations from sites described by McSorley and
Ferris (1979)

indicate that the model was very sensitive to changes in the
initial counts of P. hexincisus in the soil, since variations
in the height of the initial peak and in position of the final
peak of simulated P. hexincisus counts were observed.   Per-
formance of a sensitivity analysis before validation can point
out those areas most likely to show deviations when validation
experiments are performed.

MODEL VALIDATION

Validation involves the testing of the model against one or more independent data sets which were not involved in its derivation. Because of the sensitivity of the model to the initial soil counts of $\underline{P}$. hexincisus, it was desirable to examine several data sets covering a wide range of this parameter. Data for the validation experiments were collected from nine sites located in various parts of Indiana (McSorley 1978). Soil samples were collected at each site on the date of planting (Table 1) and assayed for initial number of Pratylenchus spp. Follow-up samples of corn roots were collected from each site between 26 and 49 days after planting, and assayed for number of Pratylenchus spp. per gram of root weight. Using the appropriate initial numbers, weather data, and agronomic data for each site, PHEX was used to provide a simulated value of the root counts on the date of root sampling. Simulated (sim P_r) and actual root counts (ac P_r) are shown (Table 1). Pratylenchus hexincisus predominated at the first six sites shown, while $\underline{P}$. scribneri Steiner occurred at site H, and $\underline{P}$. crenatus Loof occurred at sites I and O. For the six sites at which $\underline{P}$. hexincisus predominated, log (sim P_r) showed a highly significant (P = 0.01) correlation with log (ac P_r), with r = 0.931 (Fig. 6). The resulting regression line was quite close to the expected result, represented by the line Y = X. When data from all nine sites were included in the comparision (Fig. 7), log (sim P_r) still showed a significant (P = 0.05) correlation with log (ac P_r) with r = 0.767, but the correlation was lower than that obtained if points for $\underline{P}$. hexincisus alone were used. Inclusion of points from sites where other Pratylenchus species occurred caused a deviation from the expected Y = X line. Comparison of log (ac P_r) with log P_i,

TABLE 1. Comparison of Simulated and Actual Counts of _Pratylenchus_ spp. per g of Dry Root Weight for Sites Sampled in 1977.

Site	Julian Date of Planting	Initial Counts per 500cc Soil[a]	Julian Date of Root Sampling	Pratylenchus per g simulated mean	actual mean[b]
J	120	199	153	1,169	652
K	130	68	179	166	279
M	131	50	174	77	390
N	123	857	153	11,675	17,141
Q	133	378	179	9,240	8,573
R	119	279	151	2,061	1,416
H	133	370	166	2,853	345
I	140	72	166	190	15
O	132	257	166	1,097	179

[a]Mean of six replications.
[b]Mean of four replications.

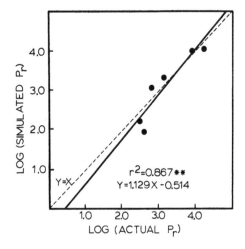

Fig. 6. Comparision of simulated and actual counts of P. hexincisus per gram of dry root weight for six sites sampled in 1977.

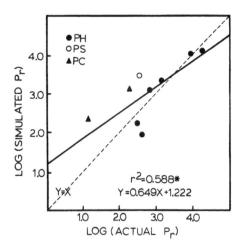

Fig. 7. Comparison of simulated and actual counts of Pratylenchus spp. per gram of dry root weight for nine sites sampled in 1977 (PH = P. hexincisus; PS = P. scribneri; PC = P. crenatus).

where P_i = initial soil counts, showed that root counts of
P. crenatus or P. scribneri were lower than those obtained
for the same initial soil population of P. hexincisus (Fig. 8).

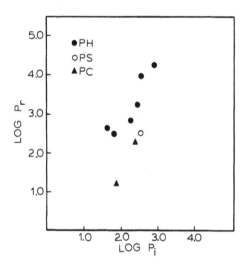

Fig. 8. Relationship between root counts (P_r) and initial
soil counts (P_i) of Pratylenchus spp. from Table 1 (PH =
P. hexincisus; PS = P. scribneri; PC = P. crenatus.

Thus, while PHEX reliably simulated early-season root counts
of P. hexincisus, consistent deviations from PHEX predictions
were observed if either P. crenatus or P. scribneri were
present in substantial numbers.

PLANT DAMAGE MODELS

The relationship between nematode numbers and plant damage
has been modeled frequently. Most of these models are static
functions relating crop yield to the logarithm of initial
nematode counts. Oostenbrink (1966) has presented a generalized model, although the expected shape of the relationship

has been a subject of some controversy (Oostenbrink 1966, Seinhorst 1965, Wallace 1973). Numerous plant damage models have been developed (Barker and Nusbaum 1971), and the methodology for the development of such relationships has been reviewed (Ferris 1978b).

ROLE OF MODELING IN PEST MANAGEMENT SYSTEMS

Nematode population models and plant damage models can be joined in online pest management systems. The structure of such systems has been described (Haynes et al. 1973), and a modified diagram is presented (Fig. 9). External inputs such as weather data, agronomic data, and sampling estimates of the field population are used by the nematode population model to generate an estimate of nematode density at some future time. The plant damage model is used to estimate crop loss at this density. Evaluation of potential damage in economic terms and comparison with costs of possible control

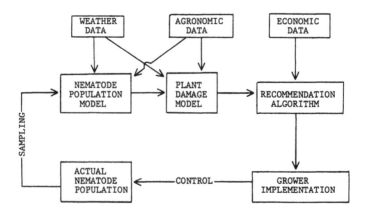

Fig. 9.   Components of a nematode management system.

strategies leads to a suggested course of action, or recom-
mendation algorithm (Ruesink 1975).   If economically justified,
a control method is then implemented by the grower.

Historically, pest management systems have evolved in an
entomological context.   During the life of the host crop,
populations of a phytophagous insect could fluctuate above
the economic injury level (EIL) several times (Fig. 10).

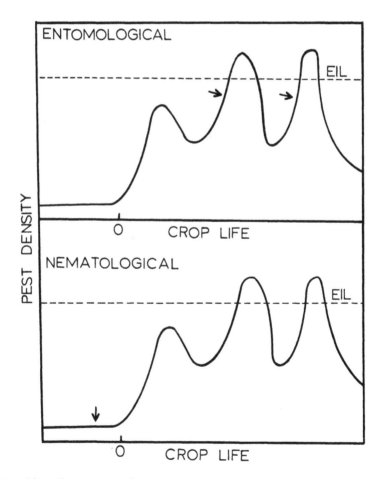

Fig. 10.   Comparison between entomological and nematological
control strategies.   Action points are indicated by arrows.

However, frequent field monitoring in a pest management program could detect these increases and a course of control could be implemented. A common control strategy is the application of an insecticide, which hopefully would bring about a rapid population decrease. Field monitoring could continue and control strategies could be implemented at several points as in Figure 10. Thus, there is a continuous cycling of information through the diagram of Figure 9. Nematological systems show a distinct difference, however, since common control strategies such as soil fumigation or application of most non-fumigant nematicides take place prior to planting. Thus, there may be only one action point in the corresponding nematological system (Fig. 10). Monitoring done after planting could have some application toward control strategies for future crops or in perennial crops, but, in general, little can be done for the current annual crop at that point except in those limited cases where registered postplant nematicides exist. The information flow of Figure 9 must then occur at one time, prior to planting. For this reason, nematode population models are needed which can project population fluctuations over the entire crop life based on preplant data.

ACKNOWLEDGMENTS
This is Journal Paper No. 8845 of the Purdue University Agricultural Experiment Station, West Lafayette, Indiana.

LITERATURE CITED
Barker, K. R., and C. J. Nusbaum. 1971. Diagnostic and advisory programs. Pages 281-306 in B. M. Zuckerman, W. F. Mai, and R. A. Rohde, editors. Plant parasitic nematodes. Volume I. Academic Press, New York, New York, USA.

Bird, G. W., M. Sarette, C. Coley, and M. Dover. 1976.
Nematology components of an on-line pest management
system. Journal of Nematology 8:280.

Ferris, H. 1976. Development of a computer-simulation
model for a plant-nematode system. Journal of Nematology
8:255-263.

Ferris, H. 1978a. Modification of a computer-simulation
model for a plant-nematode system. Journal of Nematology
10:198-201.

Ferris, H. 1978b. Nematode economic thresholds: derivation
requirements, and theoretical considerations. Journal of
Nematology 10:341-350.

Haynes, D. L., R. K. Brandenburg, and P. D. Fisher. 1973.
An environmental monitoring network for pest management
systems. Environmental Entomology 2:889-899.

Jones, F. G. W., R. A. Kempton, and J. N. Perry. 1978.
Computer simulation and population models for cyst-
nematodes (Heteroderidae:Nematoda). Nematropica 8:36-56.

Kimpinski, J., H. R. Wallace, and R. B. Cunningham. 1976.
Influence of some environmental factors on populations of
Pratylenchus minyus in wheat. Journal of Nematology
8:310-313.

McSorley, R. 1978. Components of a management program for
nematodes on corn. Ph.D. Dissertation. Purdue University,
Lafayette, Indiana, USA.

McSorley, R., and J. M. Ferris. 1979. PHEX: a simulator
of lesion nematodes in corn roots. Purdue University
Agricultural Experiment Station Research Bulletin 959.
West Lafayette, Indiana, USA.

Miles, G. E. 1974. Developing pest management models.
Purdue University Agricultural Experiment Station Bulletin
50. West Lafayette, Indiana, USA.

Mugniery, D. 1976. Etablissement d'un modele de dynamique
   de population d'Heterodera pallida Stone: applications
   a un cas pratique de lutte integree. Annales de Zoologie
   Ecologie Animale 8:315-329.

Oostenbrink, M. 1966. Major characteristics of the relation
   between nematodes and plants. Meded. Landbouwhogesch.
   Wageningen 66:1-46.

Ruesink, W. G. 1975. Analysis and modeling in pest management. Pages 353-376 in R. L. Metcalf and W. H. Luckmann,
   editors. Introduction to insect pest management. John
   Wiley and Sons, Inc., New York, New York, USA.

Ruesink, W. G. 1976. Status of the systems approach to
   pest management. Annual Review of Entomology 21:27-44.

Seinhorst, J. W. 1965. The relation between nematode density
   and damage to plants. Nematologica 11:137-154.

Wallace, H. R. 1973. Nematode ecology and plant disease.
   Crane, Russak and Company, Inc., New York, New York, USA.

A SIMULATION MODEL FOR LIFE-HISTORY STRATEGIES OF BACTERIOPHAGIC NEMATODES

Richard V. Anderson and Thomas B. Kirchner

INTRODUCTION

The use of nematodes as model systems for investigating a variety of processes from aging to nutrient cycling (Zuckerman 1980) requires a thorough understanding of the life cycle of nematodes. This has become particularly apparent in recent investigations of the effects of nematodes on mineralization of N, P, and C (Anderson et al. 1981). It has been found that nutrient dynamics are closely tied to the development stage or resistant form exhibited by the population. Nematode populations which are expanding do not return N as NH_4^+ or P as PO_4 (Cole et al. 1978 and Woods et al., in press) although they may excrete amino acids (Anderson et al., submitted). However, N and P become available in the soil solution when the nematode populations begin to decline, reverting to resistant cryptobiotic or dauer forms (Coleman et al. 1977, Anderson et al. 1981). Nematode population dynamics are also associated with CO_2 production, and C cycles in soil systems (Coleman et al. 1978, Trofymow et al., in this volume). Thus, a thorough understanding of nematode life history is necessary for interpretation of results of process oriented studies. However, as pointed out by Anderson et al. (submitted) evaluation of nutrient fluxes in relation to nematode populations may require frequent sampling since changes in population demography may occur rapidly. Processing of large numbers of samples is very time consuming and may be prohibitive in process oriented studies.

A nematode population model may solve some of the problems associated with intensive sampling by predicting population

levels. However, most of the existing models are keyed to
plant parasites and plant-nematode relationships (Ferris 1976,
Bird and Thomason 1980). Because these models have often been
directed toward pest management or development of critical
population densities, the time steps incorporated in the models
are relatively large. A high resolution model including specific
life history stages is needed in order to predict short term
population fluctuations which appear to influence nutrient
fluxes in soil systems. Critical sampling times can be determined or the interpretation of the causes of nutrient dynamics
from soil systems with known populations of nematodes can be
enhanced with this type of model in addition to obtaining a
better understanding of nematode population dynamics. The
following is a description of a high resolution model for
bacterial feeding nematodes.

GENERAL DESCRIPTION OF MODEL
The model (Fig. 1) consists of two basic submodels, carbon
flow and population dynamics. In the carbon flow model each
box or state variable represents a compartment in which
carbon (C) may be found in one form or another. In this sub-
model values are expressed in g C. A parallel set of variables and flows exist for populations and values are expressed
as numbers of individuals. The two submodels are interrelated
in that the amount of C and number of individuals reflect the
size of the individuals and weight gains or losses as a result
of food, i.e., carbon availability. Therefore the simulation
model keeps track of total biomass, total numbers and individual size at each time step.

    The core of the model represents each developmental state
during active population growth. It consists of eggs, three
distinct juvenile stages ($J_{II,III,IV}$), and adults (Croll and

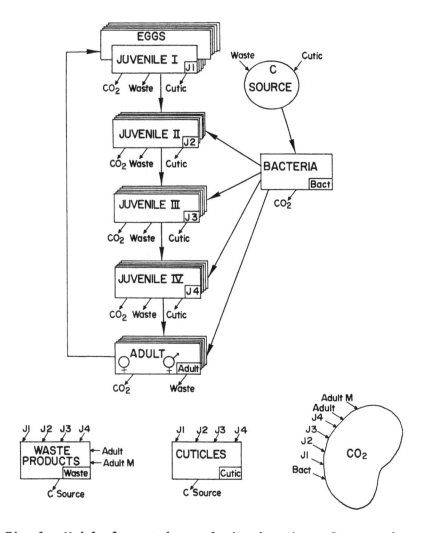

Fig. 1. Model of nematode population dynamics. State variables are developmental stages and the food source. Both
abundance and biomass - carbon are simulation. Carbon loss
from nematodes is reflected as waste products, cuticles from
ecdysis or CO_2 evolution.

Matthews 1977). The first juvenile stage frequently occurs in
the egg and thus is not distinct from that stage of the life
cycle (Croll and Matthews 1977). When the food supply (bacteria)
is abundant the nematode life cycle follows the normal pattern
of development. The rate of development of a nematode is dependent on food availability, rather than simply being an age
dependent process.

Consumption of bacteria is modeled as a simple binary
collision model of predation:

$$C = N \cdot B \cdot P_c \cdot W \tag{1}$$

where:

$\quad$ C = consumption rate in g C/h

$\quad$ N = number of nematodes

$\quad$ B = number of bacteria

$\quad P_c$ = probability of capture per nematode per
$\qquad$ hour, set at 7×10^{-4}

$\quad$ W = bacterial biomass in g C/bacterium

However, nematodes can consume bacteria only up to a rate which
allowed them to obtain a predefined maximum individual growth
rate, r_{max}. Thus, satiation of the nematodes is simulated.
The value of r_{max} can vary considerably between species (Anderson
and Coleman 1981) and needs to be determined for the species
being modeled. Some base line r_{max} values have been determined
by Yeates (1973) and Phillipson et al. (1977).

Assimilation efficiency of consumed bacteria was based on
literature values (Sohlenius 1980). This efficiency was
assumed to be constant for all stages of the life cycle with
the exception of eggs. Assimilation was 60% with 40% excreted;
of that assimilated, 38% is converted to nematode biomass
(Marchant and Nicholas 1974). Excreted carbon is transformed
to a waste compartment. It should be noted that this effi-

ciency may vary depending on type of material consumed
(Coleman et al. 1978).

The juveniles grow as they feed, i.e. biomass C increases,
until they reach a predefined size. At this point they are
transferred to the next developmental stage to simulate molting. The mass at which molting takes place will vary for
each species but can be determined from size frequency distributions based on length of individuals in a population
(Sohlenius 1973, Anderson and Coleman 1981). Length has also
been shown to be proportional to biomass (Kirchner et al. 1980)
so that biomass C for each size class can be monitored.

Most adult assimilation excluding respiration, results in
the formation of eggs. The number of eggs produced by each
reproductive adult female is a function of adult size, fecundity and longevity. These variables as well as the size of
the eggs produced are species specific. Information on egg
production in relation to adult age is supplied as a table
since egg production varies over the life span of an adult
(Fisher 1969, Sohlenius 1969, 1973, Zuckerman et al. 1971,
Popovici 1972, Anderson and Coleman 1981). The number of eggs
produced during an interval of time necessitating the inclusion
of subcompartments to follow each successive cohort may vary.
The division of each population variable throughout the model
into five subcompartments was sufficient for our simulation of
an Acrobeloides sp. and Mesodiplogaster lheritieri.

A predefined percentage of molting J_{IV} individuals would
become adult males in those species having males. It is
possible for this flow to be affected by a number of biotic
and abiotic environmental factors (Nicholas 1975). Adult
longevity is also predefined and may vary by species.

While development of juvenile stages and production of eggs
are a function of food availability, egg development is a

function of time. The food supply for the formation of the
first juvenile stage is dependent on the carbon content of
the egg. Therefore the size of the J_I stage is based on egg
size minus losses due to the egg shell and some respiration.
Although embryogenesis is somewhat variable (Croll and Matthews
1977) the development of the J_I stage was taken as 24 h based
on averages from Chuang (1962), Thomas (1965), and Croll and
Matthews (1977). This was followed by ecdysis to the J_{II} stage
which began active feeding.

The processes discussed so far involve transfers of both
numbers of individuals and biomass C. Other carbon flows include respiration, death, and the production of shed cuticles.
Respiration is determined based on the respiratory functions
of Klekowski et al. (1972), which were:

$$\mu l\ O_2 \text{ consumed} = 1.42 \text{ (live mass } \mu g) \ 0.72 \cdot 10^{-3} \quad (2)$$

the CO_2 - C evolved can be calculated from the biomass of
each developmental stage by assuming an RQ value of approximately 0.7 (Nicholas 1975). The remaining carbon flows are
associated with ecdysis. The amount of material excreted
after assimilation and adult death have already been discussed. Death during ecdysis is determined from the proportion of those individuals entering the next developmental
stage. This proportion is species specific and is specified
in the input data of the simulation model. The biomass C of
dead individuals is transferred to waste at ecdysis. Molt
mortality may be quite high, up to 30% in some species
(Anderson and Coleman 1981). The other carbon loss during
ecdysis is the shed cuticle. The proportion of the biomass
represented by cuticle is specified in the input data. Shed
cuticles are transferred to a separate state variable because the composition of the cuticle may make it more resistant

to microbial decomposition. Both the waste products and cuticle are carbon sources for bacterial growth.

VARIATIONS ON THE BASIC MODEL

Nematodes exhibit two basic life history strategies in response to adverse environmental condtions. In relation to
metabolic activity and senescence these responses can be
divided into quiescence and cryptobiosis (Cooper and Van
Gundy 1971). Because a particular species usually exhibits
only one of these types of responses each life history strategy
is modeled separately. However, the core of each model is
the same. Many environmental parameters can initiate a quiescent or cryptobiotic response; i.e. moisture level, temperature,
chemicals (Van Gundy 1965). Decreased food supply has been
shown to be an environmental factor which initiates cryptobiotic or quiescent responses in nematodes (Sohlenius 1969,
Jairajpuri and Azmi 1977, Anderson et al. 1981, Anderson and
Coleman 1981). Only the lack of food initiates the response
in the simulation model because the model is driven by availability of bacteria. The cryptobiotic or quiescent response
commences when bacteria are grazed down to a refuge level.
This level is dependent on the size of the nematodes since
their ability to access available food in small pore spaces is
limited by their ability to move through small pore necks
(Wallace 1971). The principle has been demonstrated in trophic
relationships by Elliott et al. 1980.

Cryptobiotic Response

Nematodes which undergo cryptobiosis enter that state when
bacterial populations reach a predefined refuge level (Fig. 2).
All developmental stages except eggs and the first juvenile
stage undergo this response. When bacterial populations have

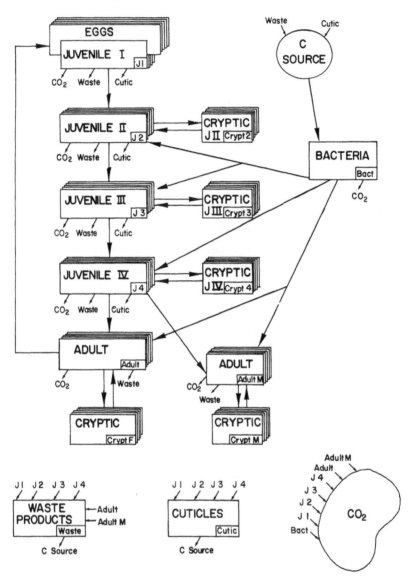

Fig. 2. Nematode population model with a cryptobiotic response to adverse environmental conditions included. There are no flows to waste, cuticle, or CO_2 because the cryptobiotic forms exhibit no measurable metabolism and do not feed.

increased to 10^7 individuals/gram substrate, the nematodes
return to the same active stage in which they were prior to
cryptobiosis and continue normal development. There is no
measurable metabolic activity (Crowe and Madin 1974) while in
the cryptobiotic state, and no CO_2 or waste products are produced. Mortality during transformation to and from the
cryptobiotic state results in a reduction in biomass-C and
number of individuals. This mortality is defined as a proportion of those individuals entering the cryptobiotic state.

Quiescent Response

This strategy is more complex because it involves both a
decrease in metabolic activity and in many cases, the maintenance of active juvenile forms (Fig. 3). Thus there are
two distinct responses, formation of "dauer larvae" and intrauterine juvenile development. Dauer larvae are quiescent forms
which develop from the J_{II} stage. The juvenile retains the
cuticle of the J_{II} stage when it develops to the J_{II} stage
(Bovien 1937, Anderson et al. 1979). Therefore the dauer
larva has a double cuticle. It is an inactive stage with reduced metabolic activity usually considered to be a form developed for dispersal, but may also be important as a method of
tolerating extremes in temperature (Yarwood and Hansen 1969)
and for resisting desiccation (Lee 1953, Evans and Perry 1975).
This response is initiated in the simulation model when bacteria reach their refuge level. When bacterial densities increase above this level the dauer larva begins to feed, molts
to a fourth stage juvenile (J_{IV}) and continues to develop
normally.

When bacterial numbers are reduced to the refuge level,
adults carrying eggs do not lay those eggs. Instead, the eggs

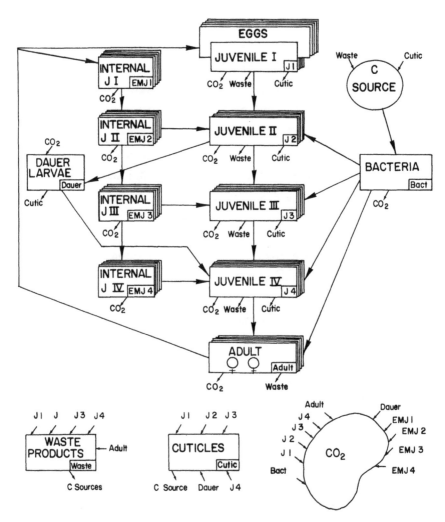

Fig. 3. Nematode population model with dauer larvae and intrauterine juvenile development. These two variations occur in response to adverse environmental conditions such as food shortage.

develop inside the adult body. The hatched juveniles feed on the decaying tissue of the adult (Otter 1933). This response occurs when food supplies are exhausted (Sohlenius 1969, Anderson and Coleman 1981) and may be a mechanism for increased juvenile survival in food-limited conditions. The response is simulated in the model by having intrauterine juveniles developing in the same fashion as normal juveniles with the exception that the food source is the female tissue. The juveniles break free of the female cuticle and continue developing after this adult biomass is completely converted to juveniles. The number of juveniles produced in this way depends on the number of eggs carried by the adult at the point when bacteria reach the refuge level.

Starvation is defined in the model as a net loss of biomass carbon due to respiration. Starvation may occur in all the nematode stages being simulated. Mortality due to starvation occurs when the biomass of a nematode falls below a minimum size. A minimum size is defined in the input data for each stage of the species being simulated. Although dauer larvae may starve to death, such starvation is much slower than that in active larvae due to the reduced metabolic activity of dauer larvae (Anderson and Coleman 1981, see also RQ values in Anderson et al. 1981).

MODEL SIMULATIONS
The model was used to simulate population development of two nematodes, each having a different type of strategy for responding to environmental stress. The two species of nematodes used were Acrobeloides sp., which has a cryptobiotic response, and Mesodiplogaster lheritieri, which produces both dauer larvae and has intrauterine larval development.

Input parameters for each nematode include minimum and
maximum size of each developmental stage. This was determined based on size frequency distributions in populations
(Fig. 4 and 5). Other parameters such as fecundity, longevity, molt mortality, etc., were from Anderson and
Coleman (1981). Values predicted by the model compared to
data from agar cultures are shown for M. lheritieri adults,
eggs and all juvenile or larval stages in Figures 6a, 6b and
6c. In M. lheritieri, dauer larvae and intrauterine egg
development began after 30 days. The larval numbers which
include the dauer forms stabilize while adult and egg numbers
decrease to zero. The model predicts this population development very well, as evidenced by a test of fit for each of the
population stages examined (Figs. 6d, 6e, and 6f). These
figures show how well the model predictions are correlated
with the experimental populations at each sampling interval.
A perfect fit would show all points lying on a line passing
through the ordinate with a slope of one. Similarly the simulation of population development of Acrobeloides sp., which
has cryptiobiotic stages, is also predicted well (Fig. 7),
with the simulation of adult and larval production fitting
more closely than that of eggs.

If the model can accurately predict population dynamics
then it can be used to evaluate C cycling in terms of nematode
contribution. Figure 8 shows model output for cuticle, waste
and CO_2 carbon produced by Acrobeloides sp. The pulses of
waste or cuticle material result from feeding activity increasing
as nematode populations develop starting with reproductive
adults. The CO_2-C response is similar to those found by
Coleman et al. (1978) and Anderson et al. (1981) for soil
cultures with nematodes.

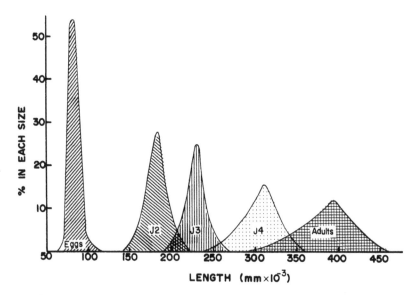

Fig. 4. Frequency distributions of size for <u>Acrobeloides</u> sp. This information is used to determine maximum and a minimum size of each developmental stage.

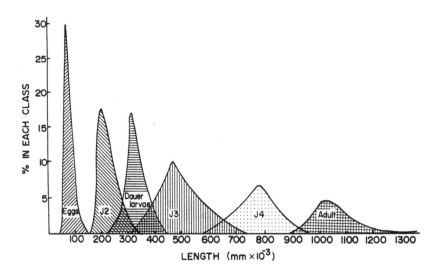

Fig. 5. Frequency distributions of size for <u>Mesodiplogaster</u> <u>lheritieri</u>. This information is used to determine maximum and minimum size for each developmental stage.

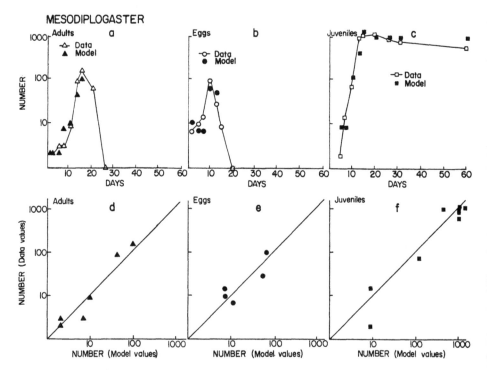

Fig. 6. Model simulation for <u>Mesodiplogaster</u> <u>lheritieri</u>, a
nematode with dauer larvae and intrauterine larval development.
a, b, and c show actual data from agar culture and abundance
values simulated by the model for adults, eggs and larvae,
respectively. d, e, and f show the relationship between labor-
tory values and predicted values of the model for adults, eggs
and larvae , respectively.

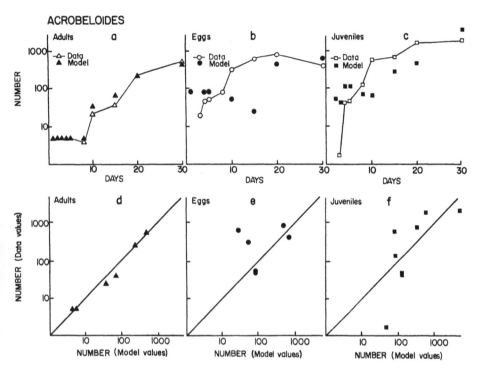

Fig. 7. Model simulation for <u>Acrobeloides</u> sp., a nematode
with a cryptobiotic response to environmental stress. a, b,
and c show actual data from agar cultures and abundance values
simulated by the model for adults, eggs, and larvae, respectively. d, e, and f show the relationship between laboratory
values and predicted values from the model for adults, eggs
and larvae, respectively.

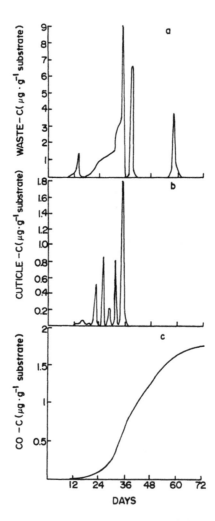

Fig. 8.   Carbon outputs in the form of: (a) waste, (b) cuticle, and (c) CO_2 evolved are plotted from a simulation using <u>Acrobeloides</u> sp.   Note the cyclic inputs of waste and cuticle carbon and the S shaped curve for CO_2 carbon.

SUMMARY AND EXTENSION

The model will simulate population dynamics of nematodes accurately. Life history strategies for responses to adverse environmental conditions have been included in the model such that food limitations affect population development. The model can be used to predict carbon flows particularly in terms of waste production, cuticle formation, CO_2 evolution, and various population parameters such as rates of maximum increase and feeding rates. The sensitivity of the population structure to changes in the rates can be determined. With the proper feeding function it would be possible to adapt the model to simulate a plant parasitic nematode. The food source could be any carbon substance given an accurate feeding function.

The model is being extended to include driving variables such as temperature, moisture, and soil texture which will affect flows between state variables. This should allow more accurate simulations of population dynamics in soils.

ACKNOWLEDGMENTS

Support of the ideas developed in this model was provided by NSF Grants DEB 78-11201 and A01 to Colorado State University.

LITERATURE CITED

Anderson, R. V., W. D. Gould, R. E. Ingham, and D. C. Coleman. 1979. A staining method for nematodes: determination of nematode resistant stages and direct counts from soil. Transactions of the American Microscopical Society 98:213-218.

Anderson, R. V., and D. C. Coleman. 1981. Population development and interactions between two species of bacteriophagic nematodes. Nematologica 27:6–19.

Anderson, R. V., D. C. Coleman, C. V. Cole, and E. T. Elliott. 1981. Effect of the nematodes Acrobeloides sp. and Mesodiplogaster lheritieri on substrate utilization and nitrogen and phosphorus mineralization in soil. Ecology 62:549–555.

Anderson, R. V., W. D. Gould, L. E. Woods, and D. C. Coleman. Nitrogen losses by microbivorous nematodes in soil. Oikos (Submitted).

Bird, G. W., and I. J. Thomason. 1980. Integrated pest management: the role of nematology. Biological Sciences 30:670–674.

Bovien, P. 1937. Some types of associations between nematodes and insects. Videnskabelige Meddelelser Fra Dansk Naturhistorisk Forening Denmark 101:1–114.

Cole, C. V., E. T. Elliott, H. W. Hunt, and D. C. Coleman. 1978. Trophic interactions in soils as they affect energy and nutrient dynamics. V. Phosphorus transformations. Microbial Ecology 4:381–387.

Coleman, D. C., C. V. Cole, R. V. Anderson, M. Blaha, M. K. Campion, M. Clarholm, E. T. Elliott, H. W. Hunt, B. Shaefer, and J. Sinclair. 1977. An analysis of rhizosphere-saprophage interactions in terrestrial ecosystems. in U. Lohm and T. Persson, editors. Soil organisms as components of ecosystems. Ecological Bulletin (Stockholm) 25:299–309.

Coleman, D. C., R. V. Anderson, C. V. Cole, E. T. Elliott, L. Woods, and M. K. Campion. 1978. Trophic interactions in soils as they affect energy and nutrient dynamics. IV. Flows of metabolic and biomass carbon. Microbial Ecology 4:373–380.

Chuang, S. H. 1962. The embryonic and post-embryonic

development of Rhabditis teres (A. Schneider). Nematologica
7:317-330.

Cooper, A. F., Jr., and S. D. Van Gundy. 1971. Senescence,
quiescence, and cryptobiosis. Pages 297-318 in B. M.
Zuckerman and W. F. Mai, editors. Plant parasitic
nematodes, Volume II. Academic Press, New York, New York,
U.S.A.

Croll, N. A., and B. E. Matthews. 1977. Biology of nematodes.
John Wiley and Sons, Inc., New York, New York, U.S.A.

Crowe, J. H., and K. A. C. Madin. 1974. Anhydrobiosis in
nematodes: evaporative water loss and survival. Journal
of Experimental Zoology 193:232-334.

Elliott, E. T., R. V. Anderson, D. C. Coleman, and C. V. Cole.
1980. Effects of soil pore space on microbial trophic
interactions. Oikos 35:327-335.

Evans, A. A. F., and R. N. Perry. 1975. Survival strategies
in nematodes. Pages 383-424 in N. A. Croll, editor. The
organization of nematodes. Academic Press, New York,
New York, U.S.A.

Ferris, H. 1976. Development of a computer simulation model
for a plant-nematode system. Journal of Nematology 8:256-
263.

Fisher, J. M. 1969. Investigations of fecundity of Aphelenchus
avenae. Nematologica 15:22-28.

Jairajpuri, M. S., and M. I. Azmi. 1977. Reproductive behavior
of Acrobeloides sp. Nematologica 23:202-212.

Kirchner, T. B., R. V. Anderson, and R. E. Ingham. 1980.
Natural selection and the distribution of nematode sizes.
Ecology 61:232-237.

Klekowski, R. Z., L. Wasilewska, and E. Paplinska. 1972.
Oxygen consumption by soil-inhabiting nematodes.
Nematologica 18:391-403.

Lee, E. 1953. An investigation into the method of dispersal
   of Panagrellus silusiae, with particular reference to its
   desiccation resistance. Journal of Helminthology 27:
   95–103.

Marchant, R., and W. L. Nicholas. 1974. An energy budget for
   the free living nematode Pelodera (Rhabditidae).
   Oecologia 16:237–252.

Nicholas, W. L. 1975. The biology of free–living nematodes.
   Clarendon Press, Oxford, England.

Otter, G. W. 1933. On the biology and life history of
   Rhabditis pellio (Nematoda). Parasitology 25:296–306.

Phillipson, J., R. Abel, J. Steel, and S. R. J. Woodell.
   1977. Nematode numbers, biomass and respiratory metabolism
   in a beech woodland –– Wytham Woods, Oxford. Oecologia
   27:141–155.

Popovici, I. 1972. Studies on the biology and population
   development of Cephalobus persegnis (Nematoda, Cephalobidae).
   Pedobiologia 13:401–409.

Sohlenius, B. 1969. Studies on the population development
   of Mesodiplogaster biformis (Nematoda, Rhabditida) in agar
   culture. Pedobiologia 9:243–253.

––––––––––––. 1973. Growth and reproduction of a nematode
   Acrobeloides sp. cultivated on agar. Oikos 24:62–72.

––––––––––––, 1980. Abundance, biomass and contribution to
   energy flow by soil nematodes in terrestrial ecosystems.
   Oikos 34:186–194.

Thomas, P. R. 1965. Biology of Acrobeles complexus Thorne,
   cultivated on agar. Nematologica 11:395–408.

Van Gundy, S. D. 1965. Factors in survival of nematodes.
   Annual Review of Phytopathology 3:43–68.

Wallace, H. R. 1971. The movement of nematodes in the external environment. Pages 201–212 in A. M. Fallis, editor.

Ecology and physiology of parasites. University of
Toronto Press, Toronto, Canada.

Woods, L. E., C. V. Cole, E. T. Elliott, R. V. Anderson, and
D. C. Coleman. Nitrogen transformations in soil as
affected by bacterial-microfaunal interactions. Soil
Biology and Biochemistry (in press).

Yarwood, E. A., and E. L. Hansen. 1969. Dauer larvae of
Caenorhabditis briggsae in axenic culture. Journal of
Nematology 1:184-189.

Yeates, G. W. 1973. Nematodes of a Danish beech forest. II.
Production estimates. Oikos 24:179-185.

Zuckerman, B. M., S. Himmelhoch, B. Nelson, J. Epstein, and
M. Kisiel. 1971. Aging in Caenorhabditis briggsae.
Nematologica 17:478-487.

Zuckerman, B. M. 1980. Nematodes as biological models,
Volumes 1 and 2. Academic Press Inc., New York, New
York, U.S.A.

SOIL SAMPLING AND PROCESSING FOR DETECTION AND QUANTIFICATION
OF NEMATODE POPULATIONS FOR ECOLOGICAL STUDIES

P. B. Goodell

In most ecological studies, whether involving energy flow or
population and community assessments, field sites or microcosms,
quantification of populations or communities is necessary. This
quantification may involve a one-time inventory or a time series
of measurements to determine patterns of change and response
to environmental stimuli. In cases where it is not feasible
to assess all individuals in the experimental universe, it is
necessary to measure a representative subset, a sample. A major
problem in the study of soil inhabiting organisms, as with other
groups of animals, is that the subset selected may not be truly
representative of the universe of interest. A sample plan must
be designed therefore, which considers the biology and ecological
requirements of the nematodes and which results in an estimate
of the population which has a measureable and acceptable level
of precision. The problem of estimating nematode numbers is
compounded by their microscopic size and soil and/or plant inhabiting nature. Quantification of the population or community
involves four steps, the collection of a representative sample
of soil or plant material, the extraction of nematodes from
this material by an appropriate technique, and the identification
and analysis of the recovered nematodes. A final and critical
step is the use of this information to estimate the parameters
of the population or community structure of nematodes in the
sampled universe, a calculation which inherently requires knowledge of the errors and variability associated with each of
the foregoing steps.

The process of quantifying a nematode community may be considered as a funnel (Fig. 1) which reduces and condenses the

sampled universe. The reduction in complexity includes a
subsequent reduction in information which is accepted by the
investigator in the budgeting of time and resources. The loss
of information and error associated with each step must be
quantified and thoroughly understood by the investigator since
the subsequent magnification process in estimating the parameters
of the population will compound errors and may result in unrep-
resentative and biased estimates. The objective of this presentation is to introduce the basic procedures of the process,
to discuss their effects on reduction of information and
associated error and to explore the components of the process
which are of a major importance in designing a sampling strategy
for ecological studies.

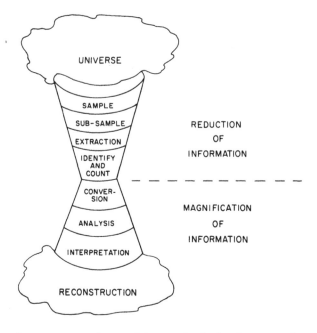

Fig. 1.   Components of sampling and their interactions.

SAMPLING OF SOIL AND PLANT MATERIALS

There are several major factors to consider in designing a
sampling strategy. The biological reality of the situation is
perhaps the most important factor to consider. It consists of
all of the factors in the environment which affect, directly
or indirectly, the spatial or temporal distribution of the
nematode. The nematode fauna is generally diverse and is involved in complex food webs at all trophic levels. Biology of
the nematode is a prime determinant of the time, the place and
depth of the sample.

Since plant roots are the primary source of energy in the
soil ecosystem, the roots and associated rhizospheres have a
major effect on the distribution of nematodes in the soil. The
axes of the distribution model for soil inhabiting nematodes
are vertical, horizontal and temporal. Among the factors influencing distribution are rooting pattern of the primary pro-
ducers in the system, soil texture, temperature, moisture content
and oxygen availability and possibly pH. In systems undergoing
rapid or frequent vegetative change, the structure and host
status of any previous plant community is important.

Nematode distribution can be considered at two levels; the
micro-distribution of the individuals around a single root system,
and the macro-distribution of the population over a larger area.
Micro-distribution of nematodes at the primary consumer level
will be determined primarily by the feeding habits of the
nematode on the root system (root tip feeders, root hair feeders, corticaccal feeders, vascular feeders, etc.) and its reproduction or ovipositing habits (eggs deposited individually in
soil or aggregated in masses). Also involved are tolerance of
the nematode to the various physical factors in the micro-en-
vironment and the relationship of these factors to soil depth.
Nematodes at other trophic levels are also distributed largely

according to their food sources, but again, within the range
of suitable physical environmental conditions. Macro-distribution of nematodes is determined by soil type and textural
differences, drainage patterns, perhaps slope and aspect, vegetative composition and land use or management patterns.

In a study of population dynamics of several plant-parasitic
nematodes in a vineyard, Ferris and McKenry (1974) found substantial differences between two species. Xiphinema americanum
mainly occupied the upper 30 cm of undisturbed soil within the
vine row (Fig. 2), while Meloidogyne arenaria had high populations throughout the depths sampled including the disturbed
region between vine rows (Fig. 3). Numbers of Meloidogyne
showed greater fluctuations than those of Xiphinema throughout
the year. Reproductive strategies and biological requirements
of each species may account for these differences. Xiphinema
is a large ectoparasitic nematode with an associated activity
level requiring relatively high oxygen levels, and is less
tolerant of soil disturbance than Meloidogyne, a sedentary
parasite. Freckman and Mankau (in press) demonstrated the influence of the rooting patterns of desert shrubs on the distribution of associated nematodes at all trophic levels.

Variation in macro-distribution of nematodes is illustrated
by a study of a 7 ha alfalfa field in which individual soil
cores on a 6 m grid pattern were removed and the nematode
communities assessed (Goodell and Ferris 1980). Distributional
patterns varied with individual species, even though, all those
considered were plant parasites and feeding on a common host.
There are great differences between nematode counts 6 m apart
(Fig. 4), presumably reflective of nematode biology and host
root distribution. On a larger scale, distribution was influenced by a fine-textured band of soil across the field
(Fig. 4a). Helicotylenchus digonicus (a migratory ectoparasite)

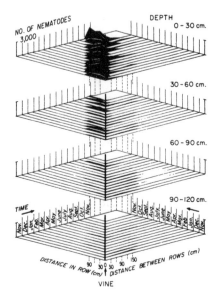

Fig. 2. Distribution of <u>Xiphinema americanum</u> in a California vineyard. After Ferris and McKenry, 1974.

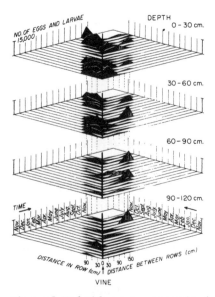

Fig. 3. Distribution of <u>Meloidogyne arenaria</u> in a California vineyard. After Ferris and McKenry, 1974.

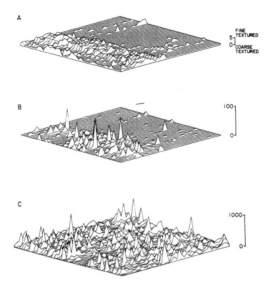

Fig. 4. Distribution of (a) soil texture, (b) Helicotylenchus
digonicus, and (c) Meloidogyne arenaria in an alfalfa field
in California. After Goodell and Ferris, 1980.

was correlated with the fine-textured area (Fig. 4b) while
Meloidogyne arenaria was generally distributed but more
prevalent in the sandy areas (Fig. 4c).

As with other nematode distribution studies (Anscombe 1950),
Merny 1970, Proctor and Marks 1975), the frequency of occur-
rence of nematodes in individual samples from the alfalfa field
study was adequately described by a negative binomial distribution. This distribution is indicative of aggregation or
clumping of the population, reflective of the patterns of
micro- and macro-distribution associated with nematode biology
and food sources already discussed. This aggregation places
constraints upon the precision which can be achieved by any

sampling strategy and upon the amount of effort necessary to
achieve that precision. Population estimates of organisms
with an aggregated distribution have a variance which is related to the mean and which is greater than the mean.   This
indicates that there will always be a considerable level of
uncertainty associated with any estimate of population density
from a sample. One of the parameters of the negative binomial
distribution descriptive of the degree of aggregation is the
k value or index of dispersion (Elliott 1971, Southwood
1975). The higher the k value, the more the population is
dispersed, until a situation is reached where it is no longer
considered aggregated and approaches a random distribution.
Indices of dispersion (k) for the nematode populations in the
alfalfa study (Goodell and Ferris 1980) were extremely low,
generally in the range of 1-2, and varied with species.

Once the negative binomial distribution has been confirmed
for a population of interest, and the index of dispersion determined, it is possible to calculate the number of samples which
should be taken to estimate the population density with a
specified level of precision (Elliott 1971, Southwood 1975).
This allows the investigator to understand and to state the
precision of estimate of the population data which, if interested
in energy flow consideration, he may be using as a base for a
multitude of subsequent calculations. The clumped distribution,
with the relationship between the mean and variance, also
places restrictions on the use of certain statistical tests
which require that the variance be normally and independently
distributed. Consequently, transformation of the data may be
necessary to avoid violation of the basic assumptions of such
parametric tests.

The time of year in which the sample is taken may greatly
influence the population estimate, as illustrated by the vine-

yard studies (Fig. 2,3). Barker et al. (1969) demonstrated
that seasonal fluctuations varied between species according to
their biology, life styles, population life stage structure
and activity levels (Fig. 5a, b).

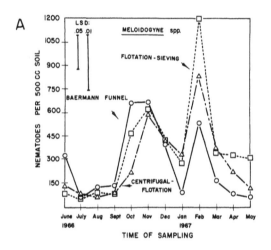

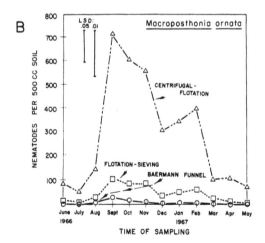

Fig. 5. The seasonal fluctuation of (a) <u>Meloidogyne</u> sp. and
(b) <u>Macroposthonia ornata</u> as measured by three extraction
methods. After Barker et al., 1969.

The basic knowledge of the organism is necessary to allow choice of a single sampling date as representative of that population, and it may be necessary to sample at intervals throughout the year to determine seasonal fluctuations. Nematodes are, of course, poikilothermic organisms and, as such, their physiological activity is closely related to the ambient temperature conditions. Consequently, in a seasonal population study it may be possible to space sampling intervals according to a physiological time sequence (degree days above a threshold) to allow increased sampling intensity when the nematodes are physiologically active and decreased intensity when they are inactive.

Another major determinant of the design of a sample program are the objectives of the study. It is essential that these objectives be clearly defined by the investigator prior to designing the sampling program. For example, an investigation conducted to determine the presence of a specific nematode may require a more intensive sample than in an investigation designed to enumerate population densities. The required levels of accuracy and estimates of precision for the two objectives would also differ.

The confidence levels associated with the sampling strategy are the levels of accuracy and reliability of the program. They may be limited by the resources available to the investigation and may be limiting to the possibility of achieving the stated objectives. A preliminary decision must be made on the level of accuracy necessary to distinguish between true population differences and sampling error. Replication of the sample is necessary to determine the variance between sites and dates, and to allow statistical evaluation of differences. Again, it is possible once the nature and parameters of nematode distribution have been defined, to determine the number of samples

necessary to estimate the nematode population density with a
defined level of confidence (Elliott 1971, Southwood 1975).

In consideration of the factors outlined, a sample design
is formulated. Components of this design include the sample
unit, the sample size, the number of samples, the technique
to be used, and the economics of the process. Generally a
soil or plant sample taken for nematode analysis consists of a
collection of non-overlapping sample units which should cover
the entire population (Cochran 1977). There are many potential
sample units in nematode assessment, as represented by the
various tools available for taking samples (Fig. 6). The
sampling tool must remain the same during the entire study to

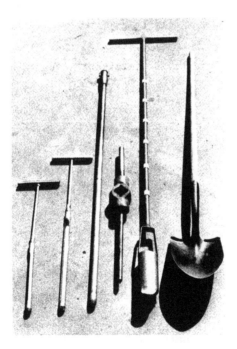

Fig. 6. An example of some nematode sampling tools. Left to
right 12" Oakfield tube, 18" Oakfield tube, Veihmeyer tube and
hammer, auger bucket, shovel.

avoid introducing a large bias. Andrassy (1962) studied the
influence of increasing sample unit size on the enumeration of
nematode species in forest litter and determined a unit size
beyond which there was no further increase in the number of
species counted. Such studies are useful preliminary deter-
minates of the appropriate tool for a particular investigation.
A frequently used method for soil sampling is to collect a
number of soil cores using a 1" diam tube inserted into the
soil to a depth of 12 to 18 inches. Depending on the study,
these cores may be taken from the root zone of specific plants
and from particular regions of the root zones of a plant to
avoid compounding sampling error with micro-distributional
variation. The size of the sample then is a function of the
predetermined number and appropriate size of sample units.

The depth of the sample is an important factor to be considered in the sample design. Again, this consideration will
be reflective of the target nematode biology and distribution.
It will also determine the sampling tool to be used. A 1-inch
diam tube can only be pushed into most soils to a depth of
about 18 inches. However, it may be possible to remove the
upper 18 inch core and, using handle extensions on the tube,
to collect the second 18 inch depths provided the walls of the
sample hole do not collapse. In other cases it is necessary to
use a sturdier tube driven into the ground by a heavy hammer or
an auger which is drilled into the ground and soil removed at
depth increments (Fig. 6). These latter approaches are labor-
intensive, and may result in a reduction in the number of sample
units due to resource constraints and consequently a reduction
in the precision of the population estimate.

The more cores or units included in a sample, the greater
the expected precision of the population estimate. However,
because of the relationship between the variance and mean in

aggregated populations, there is clearly a point of diminishing returns beyond which the increased precision is not justified by the increased cost. In the sample optimization study (Goodell and Ferris 1981), as sample size (number of cores per sample) or sample number was increased, deviation of the estimate from the true mean of the entire field decreased (Fig. 7). However, the decrease in the deviation was not linear but rather exponential, decreasing rapidly with the initial increase in sample size or

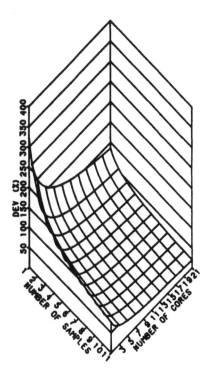

Fig. 7. The influence of increasing sample size and number on the deviation from the true mean of population estimates of Helicotylenchus digonicus in a 7 ha alfalfa field in California. After Goodell and Ferris, 1981.

number but gradually leveling off. Usually, the cost of increasing sample number or size must be considered since the available resources are rarely unlimited. As the number of samples is increased, so the number of extractions and counts associated with these samples is increased. As sample size increases by accumulating greater numbers of cores, there may not be an increase in the number of subsequent extractions and countings. Consequently, the cost of increasing sample size by bulking together greater numbers of cores is generally less expensive than increasing number of samples (Fig. 8).

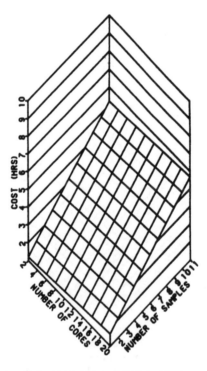

Fig. 8.   Influence of increasing sample size and number on the cost of sampling. Cost includes collecting, processing, and counting the samples. After Goodell and Ferris, 1981.

Another major consideration in experimental studies involving soil sampling is that the sampling process itself is destructive to the study site. A single soil core thrust into a plant root system may do more damage to the plant than many thousands of nematodes. Similarly, it may have a severe microhabitat impact. Consequently, the necessary number of samples required to achieve a certain level of precision may then become a determinant of plot size in experimental work.

If the sampling site has some biotic or edaphic variations which could reasonably be expected to influence nematode biology and distribution, it is justifiable and advisable to stratify the area. Then the sampling strategy would be to remove samples from each individual stratum in order to obtain an estimate for that area, and also an indication of the differences between areas. The objective in stratification is to minimize the variance among repeated samples taken from a single stratum while maximizing variation between strata (Cochran 1977). An additional benefit of this approach may be the mapping of the distribution of the nematode community.

A preliminary sample survey may be necessary to determine the design of the sampling program, its practicality and its associated variance. This allows the investigator to determine if, within the available resources, population estimates can be made to determine population densities with sufficient precision to satisfy the objectives of the study. The result of such a survey may force the redesigning of the experiment or redefining of the experimental objectives.

A final note on collecting soil samples. Considering the time and energy invested in the collection, and the fact that any estimate of the nematode population is only as good as the sample from which it was taken, it is essential that samples be treated with extreme care prior to processing. Many extrac-

tion techniques, and frequently the identification of nematodes,
requires that they be living and in a healthy condition. It is
important to keep soil samples out of the sun and to transport
them to the laboratory under a state of mild refrigeration
(15°C).

EXTRACTION OF NEMATODES FROM SOIL SAMPLES AND PLANT MATERIAL

There are several excellent descriptions of nematode extraction
processes and of extraction efficiency (Thorne 1961, Barker et
al. 1969, Ayoub 1980, Freckman and Baldwin, in press). Consequently, a detailed consideration would be repetitive. Rather
a general overview will be attempted and some important considerations emphasized. There are two basic processes by which
nematodes are removed from soil or plant material. Either they
are encouraged to move out themselves through their own motility,
or they are floated out by means of an upward current of water
or flotation in a dense solution. All extraction techniques are
based on either one of these methods, or a combination of the
two. All methods have problems and it is important to select
an appropriate method for the types of nematode anticipated in
the experimental site and for the objectives of the study. Extraction procedures vary selectively in their efficiency for
the recovery of various nematode species. Extraction efficiencies
for most methods are notoriously low (in the region of 10-25%
recovery), and may also be extremely variable and subject to
operator error. As with sampling, it is essential that the
errors and inefficiencies associated with the technique selected
be quantified to allow a true assessment of the nematode population under investigation. Communication of numbers and biomass, and consideration of energy flow through ecosystems, are
meaningless unless extraction efficiencies have been taken into
account.

Representative of techniques in which nematodes move from
the sample by their motility is the Baermann funnel. At its
simplest, a quantity of soil is placed on a tissue supported
by a screen in the top of a funnel filled with water. The stem
of the funnel is closed by a tube and clamp so that nematodes
collect as they move through the tissue and gravitate to the
base of the funnel. After a prescribed time period (usually
48 h) the clamp can be released and nematodes collected from
the stem of the funnel. Advantages of this method are the low
cost and portability of the equipment, while the disadvantages
are the dependence upon nematode motility which may differ
among species or life stages, and which may have been affected
since the sample was removed from the soil.

In the decanting and sieving process, a sample is suspended
in a large volume of water, stirred thoroughly, heavier particles are allowed to settle and the suspension poured through a
series of sieves. A water slurry containing the nematodes can
be washed from the finer sieves and the nematodes removed on a
Baermann funnel or by a centrifugation process. The sieving
procedure is the limiting step in this process and the one
which determines extraction efficiency. A sieve opening
larger than the diameter of the nematodes will result in
nematode loss unless the nematode falls horizontally across
the sieve mesh. The percentage loss will vary with the size
and shape of the nematodes. If the sieve openings are too
small, the mesh will clog with fine soil particles, resulting
in overflow and loss of the sample. The efficiency of the
sieving and decanting process may vary with soil texture. It
is necessary to define the extraction efficiency for each size
category of nematode present in the study area, and for any
variation in soil texture. If there is a substantial loss
through the sieve, it may be reduced by using a series of sieves

of equal mesh. The assumption here is that if there is a percentage loss through a sieve, then some of those passing through
the first sieve will be caught in the second sieve, and so on.
If the percentage loss at each sieving is known, the number of
sieves to be stacked to provide a certain extraction efficiency
can be calculated.

The centrifugation method employs the density differential
between nematodes and soil particles. A sample suspended in a
volume of water and decanted over a series of screens is washed
into a centrifuge and suspended in a 1 Molar sugar solution.
During centrifugation, the soil particles settle while the
nematodes remain in suspension. The solution is decanted onto
a fine mesh sieve, backwashed into a tube, suspended in water
and centrifuged again. The nematodes are then in a clump at
the bottom of the tube, and can be removed into a dish for
counting. This method is efficient in extracting certain
nematode species, but is limited by the efficiency of the
preceding decanting and sieving technique

Time of year, age structure of nematode populations, and
soil texture are major factors influencing nematode recovery
efficiencies. Further, extraction techniques differ in efficiency for different nematode genera (Fig. 5) (Barker et al.
1969). The age structure of the population may be such that
most of the population is in an egg stage or a non-motile
sedentary adult stage. In this case, extraction techniques
requiring nematode motility would provide a biased estimate of
the population. Finer textured soils are frequently more difficult to work with in nematode extractions, since they tend to
block holes in sieves and result in overflow. Further, aggre-
gations of particles may trap nematodes and prevent their
recovery.

Nematodes are usually extracted from plant tissue in a mist chamber extraction system in which roots or plant tissue are suspended in a screen over an open funnel, the stem of which is suspended in a test tube of water. The apparatus is placed in a mist chamber so that a warm mist bathes the root. Those nematodes which are still in an active and motile stage move to the surface of the root, are washed off by the mist, and settle to the base of the test tube. Extra moisture gathering in the funnel also gravitates to the test tube, but sufficiently slowly to overflow at the rim without disturbing the nematodes settling at the base. In most mist chamber extractions, there is a certain loss of nematodes in the overflow of the excess water, varying with the type and size of the nematode. Further, some nematodes within plant tissue are non-motile and cannot be recovered. Alternative assessment methods such as staining of root tissues and microscopic assessment will be necessary.

Prior to commencing a study, preliminary testing techniques are recommended. It is important that the extraction efficiency of the selected techniques for the nematodes to be recovered are determined. Care should be taken for consistency in the use of the technique so that the estimates from different samples are comparable. Extraction efficiencies can be determined by seeding sterile soil with known numbers of nematodes, and then determining percent recovery. This allows the sample estimate to be expressed in absolute terms and allows comparison of estimates made by different techniques.

IDENTIFICATION AND COUNTING

The importance of proper identification of nematode species re-covered from soil samples is obvious, but cannot be understated. Mis-identification of the nematodes may result in incorrect perception of their role and niche in the ecosystem, and thus

invalidate the whole study. To obtain help with nematode identification it may be necessary to send specimens to individuals
specializing in various nematode groups. There are prescribed
methods for preserving, mounting and shipping material for
identification by a cooperating expert (Ayoub 1980).

GENERAL SUMMARY OF RECOMMENDATIONS
Studies aimed at the quantification of any part of an ecosystem
require sound sampling plans. Results from even the most well-
conducted investigations are meaningless if the variance of the
sample universe is unknown. Sound sampling plans incorporate
well-defined objectives, knowledge of the nematodes temporal
and seasonal distribution and of its biology, acceptable levels
of accuracy and reliability, and still remain economically
feasible. Methods are available for conducting preliminary
sampling surveys of a study site to determine the appropriate
sample strategy and its associated precision. The biology of
the nematode influences the timing and depth of the sample
(seasonal or daily vertical fluctuations, seasonal population
dynamics and changes in population structure), choice of the
sampled medium (soil, plant or water), and choice of an appropriate extraction system (Barker and Campbell 1981). The
efficiencies and errors associated with the extraction system
should also be determined and considered in the estimate of the
nematode population density in the sample universe. The
clumped distribution of nematodes must be considered in
selecting sample depth and location, and in analysis of results
(transformation of data or the use of nonparametric statistics).
Sample variance can be reduced by increasing the number of
samples or by increasing sample size (bulking soil cores
together). Of course, a point of diminishing return is reached
between increase in sample size and number and increase in the

precision of the estimate. Optimization studies with plant-
parasitic nematodes in agricultural situations indicate that
this point is at about 12 to 20 cores per sample for a 1-inch
diam sample tube (Goodell and Ferris 1980). Considerable work
is necessary in this area, and ecological studies should always
be preceded by an analysis of the errors associated with the
techniques to be used.

LITERATURE CITED

Andrassy, I. 1962. The problem of number and size of sampling
unit in quantitative studies of soil nematodes. Pages 65-67
in P. W. Murphy, editor. Progress in soil zoology.
Butterworth's, London, England.

Anscombe, F. J. 1950. Soil sampling for potato root eelworm
cysts. Annals of Applied Biology 37:286-295.

Ayoub, S. M. 1980. Plant nematology. An agricultural training
aid. NemaAid Publications, Sacramento, California, U.S.A.

Barker, K. R., C. J. Nusbaum, and L. A. Nelson. 1969. Seasonal
population dynamics of selected plant-parasitic nematodes
as measured by three extraction procedures. Journal of
Nematology 1:232-239.

Barker, K. R., and C. L. Campbell. 1981. Sampling nematode
populations. Pages 451-474 in B. M. Zuckerman and R. A.
Rohde, editors. Plant parasitic nematodes. Volume III.
Academic Press, New York, New York, U.S.A.

Cochran, W. G. 1977. Sampling techniques. Third Edition
John Wiley and Sons, New York, New York, U.S.A.

Elliot, J. M. 1971. Some methods for the statistical analysis
of benthic invertebrates. Freshwater biological association.
Scientific Publication Number 15, Ambleside, England.

Ferris, H., and M. V. McKenry. 1974. Seasonal fluctuation in
spatial distribution of nematode populations in a California
vineyard. Journal of Nematology 6:203-210.

Freckman, D. W., and J. G. Baldwin. Soil nematoda. in: D. L.
  Dindal, editor. Soil biology guide. John Wiley and Sons,
  New York, New York, U.S.A.

Freckman, D. W. and R. Mankau. Distribution, abundance and
  productivity of soil nematodes in a northern Mojave desert
  site. Journal of Arid Environments. (in press).

Goodell, P., and H. Ferris. 1980. Plant parasitic nematode
  distribution in an alfalfa field. Journal of Nematology
  12:136-141.

Goodell, P. B., and H. Ferris. 1981. Sample optimization for
  five plant-parasitic nematodes in an alfalfa field.
  Journal of Nematology 13:304-313.

Merny, G., and J. DeJardin. 1970. Les nematodes phytoparasites
  des rizieres inondee de Cote d'Ivoire. II. Essai d'estimation
  de l'importance des populations. Cah. Orstom, ser. Biol.
  11:45-67.

Proctor, J. R., and C. F. Marks. 1975. The determination of
  normalizing transformations for nematode count data from
  soil samples and of efficient sampling schemes. Nematologica
  20:395-406.

Southwood, T. R. E. 1975. Ecological methods. Butler and
  Tanner, Ltd., London, England.

Thorne, G. 1961. Principles of nematology. McGraw-Hill Book
  Co., Inc. New York, New York, U.S.A.

DISCUSSION

Ferris summarized the papers which had been presented and
opened the session for discussion.

DUNCAN: I have a question for Anderson concerning his
model. Do you intend to put the plant growth component into
your model?

ANDERSON: I think you would probably have to, as a carbon
source.

DUNCAN: Do you know of any model that would be suitable for
that or do you plan to develop one?

ANDERSON: I would have to turn that over to Ingham.

INGHAM: The conversion to the plant feeding forms is in its
conceptual stage right now. My personal feeling is that we
can generate our own plant model and hopefully keep it simple
enough.

YEATES: I noted when I had regressions of nematode numbers
in various areas that I could get meaningful results only if
I used one person's data, one person's extraction methods
and errors. Now we have a series of people compiling models
for various things and those models are going to have inherent in them biases which were obtained in their development.
However, if you validate the system and it is sensitive
enough, you will get all the errors and biases.

WHITFORD: That would depend on the kind of data you use to
validate the system. If you develop a model based on microcosms, you should use something completely different and independent of a microcosm for validation. Then it would seem
to me that even though variation occurs between investigators,
if the model has some biological reality you can put your

finger on the single processes. I think the other thing that we need to do when we have enough information is to construct a model, and use the model to help direct our research. This will help us ask more robust questions and design more robust experiments. Dave Coleman and I have been talking about the two approaches; the one I use, which is taking an axe to the system and the one Dave uses, which is using microsurgery or his microcosms. They are both very excellent approaches to get to different kinds of things, but what you learn with microsurgery you need to test in the field. And then you can also use the microcosms to evaluate whatever agents you are using to help you interpret your field results.

GOODELL: I would like to say that in respect to the plant models that it is difficult enough trying to develop a nematode model or a zoological model, and if we wish to build valid plant models, we had better bring in some agronomists and work closely with them, although I know there is not a lot of that being done. I know a lot of meteorological monitoring needs to be done and fine levels of monitoring should go into that model which we don't normally collect, or even have any desire to collect.

WHITFORD: There are, in fact, plant physiologists and ecologists developing plant growth models which might be readily converted to mimicking the plant crop system.

GOODELL: I think it might be a lot easier to go in that direction and have a separate plant growth model.

WHITFORD: Some of these models have already been built and they are physiological models. It would be a matter of knowing enough about the physiology of the crop plant that you want to model and it would be my guess that there is a lot more known about the physiology of the crop plants than there are about some of the other plants. If enough of those models exist, you should be able to interest somebody into converting those models.

ANDERSON: There are some of these models available, although very few people know about them. Last spring, I went to the crop simulation workshop, which has been going on for a number of years at Mississippi State University and their complaint was that the entomologists are building all their models and very few of these will tie in with the crop models, so there exists a real communication problem.

WHITFORD: That was the same sort of complaint that came out of the soybean conference. There they had models of nutrients, photosynthesis and growth and carbon allocation and no one uses the models.

COLEMAN: I would like to point out that even though there has been quite a lot of progress in the agronomic crop modeling side of things, that in the past, even getting a simple sort of model of root uptake or nutrient uptake that included the soil fauna, and not to mention the soil microflora, has generally been accepted as an uninteresting side effect. Whereas I am sure that these types of interactions would be able to inform us as to why cotton grew so well in one type of field and maybe not so well in another type of field. So I don't know if this is just a plea for more of this treatment interaction or an exchange of ideas!

FERRIS: One of the things that has concerned me about the development of a root model, from the standpoint of plant parasitic nematodes, at least, is that I can't really conceive of them as being biomass increasers or whatever, and there would have to be some sort of root morphology built in the model. We are interested in numbers of root tips, infection sites, and damage per certain parts of the plant and that becomes somewhat difficult.

WHITFORD: With regard to the plant parasitic nematodes, I think it would be very useful to examine the other fauna that may be acting as predators on the eggs, larvae, or whatever life stages of the plant parasitic nematode that are outside the root, because someone pointed out today that 90% of the eggs produced never hatch.

FERRIS: Or don't result in a reinfection.

WHITFORD: Or result in a reinfection that you can measure. So something is happening there and if you change that mortality just a little, you can certainly, it would seem to me, change the magnitude of the impact that population is going to have on the plant whether it is a crop plant or a native plant growing in a prairie. Even though the nematodes are parasitic, there are things that prey on them. Our crop control practices and our pest control practices may be doing the same thing that I found in the desert in my experiments, and that is knocking out some fauna which may potentially control the population.

FERRIS: Well, I think that 90% egg mortality, and that's what the figure is, is a real reflection of our ignorance of the biological suppression. We have those kinds of measurements and it's been only the last 2-3 years that there has been an increasing awareness of egg parasitism on plant parasitic nematodes. In the Imperial Valley of California, in sugarbeets, the cysts of the sugarbeet cyst nematode are 80% parasitized and so the measurements which are made on economic threshold, etc. are based on these infected eggs. In central California, we located some peach orchards where we could see evidence of the presence of root-knot nematodes, but there were very few larvae in the soil although the roots were galled and the infection was there. The peaches had been there for 30-40 years and this is the ideal condition for root-knot nematode development. When Stirling looked in these orchards, we found there was a fungal egg parasite of root-knot, so there was a very low percentage of egg hatch. Next to this peach orchard, there was a grape vineyard with the same type of soil texture, of about the same age, same nematode population, but with millions of root-knot nematodes and no egg parasites. So, here is an example of something either being done in the cultural practices of that peach orchard vs the grape vineyard, or perhaps a plant input which made a favorable environment for the nematode survival.

YEATES: Your information on egg mortality refers to the work that Graham Stirling was working on?

FERRIS: No, that's just a general situation. We have a pot in the greenhouse that we add a known number of nematodes to and then we stain the roots and find out how many were infective, and it's usually a very low percentage.

YEATES: So it's 10% of the infective juveniles that have gotten into the roots.

FERRIS: Yes. And it varies, although there is some experimental error involved in that.

YEATES: In the field, of course, we do get differences with invasion of _Heterodera_ where we are talking about mortalities of say 20% or 80%, a difference seen in autumn and spring, depending on soil conditions.

FERRIS: And that could be direct environmental effect on the nematode or some indirect effect on the biological antagonist.

YEATES: But every spring is the same and every autumn is the same.

FERRIS: Dr. Coleman, in your studies in Colorado, are you validating some of the models you are getting, taking them out into the field, and seeing what's happening in the

prairies, or is the validation more at the microcosm level?
What are the objectives of those microcosm studies? Do they
end up showing what is happening to the prairie?

COLEMAN: It is more of a basic process study in which we
would like to work our way into the field oriented studies.
Possibly, either Tom Kirchner or Rick Anderson would like to
comment on this. If we know something about the biology of
the nematode, that is, if we know it goes through a series of
larval stages, and we know that under certain adverse conditions, it will go into a cryptobiotic stage, then all we were
really interested in was models of basic biology and using it
to simulate population fluctuations. And, in fact, in the
sort of response that Gregor has seen, we really did not use
microcosm populations to produce the model. We used the
known biology of the nematode and then tried to see if the
model simulated the populations that we got in microcosms.
The real purpose of that model was to predict population
ranges of a nematode and not necessarily the ones that we
use. Given the input data for that model, we should be able
to predict populations of any nematode population response,
but to add a qualifier to that, it requires a fair amount of
parameters and we need to know things like the mass and size
of the various larvae, biomass, and feeding responses. There
are a whole set of things which need to be input and that's
a problem. We are fortunate that we have some of the data for
a few nematodes we have worked intensively with and we can
simulate the model in microcosm studies, but the data does not
appear to be available for other nematodes.

DUNCAN: Presumably it should be more available for plant
parasitic nematodes and that will be one of the factors that's
going to drive the model.

COLEMAN: And this has an impact and subsequent response
with the difference between bacterial and plant feeding nematodes. Work at the detailed physiological level has not been
done. In a recent review there were only a few papers where
you could guess at an assimilation efficiency and this gave
a figure of 100% efficiency, so in fact, you don't have much
information but doing simulation models like this does help
you to hone in on the key areas.

ANDERSON: Going back to the assimilation efficiency, for my
work, it does appear that food quality has a lot to do with
assimilation efficiency. For instance, some of these omnivores that may be feeding on other nematodes appear to assim-
ilate that material much more efficiently than bacteria and
perhaps similar responses would be seen in plant tissues.

FERRIS: Well, I can think of root-knot nematodes with
modified feeding sites and a lack of desire to poison their
root environment by shunting out all kinds of waste. They
are perhaps more efficient in that manner than some of the
ectoparasites that are feeding on unmodified tissues.

GOODELL: With respect to that, you would expect that a
parasite, such as root-knot, to be so in tune with the plant,
that when the plant starts to crank down, as the plant goes
into senescence for the winter, the nematode would be willing
to crank down and go into some sort of survival stage. I
don't recall what you were saying about your high mortality,
Gregor, but was that the spring or the fall that you found
your egg death?

YEATES: That was in the spring, I think. And it is apparently due to water logging and that means it isn't due to
fungal attack.

GOODELL: I could just imagine that root-knot would be very

attuned to the changes in the hormone balance of the plant
and as the plant slows down for the winter, the nematode responds by reducing the number of eggs.

KIRCHNER: Well, that was one of the reasons I was asking
about nitrogen earlier. Translocation and nitrogen in plants
follow a very common path throughout the leaves before they
senesce and that's good stimulus for the production of protein. But I also know aphids have trouble when they are with-
drawing a lot of photosynthate or sap. They have to proccess
a lot of that to get the nitrogen to build tissue. I was
wondering if root-knot nematodes in selective feeding sites
have to process a lot of carbon from photosynthetic materials
to get a lot of protein and pump that out which would increase
the energy flow.

FERRIS: I don't have any data on that. Obviously, it is
necessary in the study of the role of nematodes in energy
flow through ecosystems to know more about what it is they eat
and how specific they are, about what they will eat, and when
they will eat it and so on. These types of studies are
necessary and should be pursued on a large scale.